A New Technique for
Job Analysis

A New Technique for Job Analysis

Walter Rohmert and
Kurt Landau

Taylor & Francis Ltd
London and New York

Originally published 1979 by Verlag Hans Huber, Bern,
Stuttgart and Vienna under the title Das Arbeitswissenschaftliche Erhebungsverfahren zur Tätigkeitsanalyse (AET)
© 1979 Verlag Hans Huber

English edition published 1983 by Taylor & Francis Ltd,
4 John Street, London WC1N 2ET
© 1983 Taylor & Francis Ltd

All rights reserved. No part of this publication may be
reproduced, stored in a retrieval system, or transmitted in
any form or by any means, electronic, mechanical, photo-
copying, recording or otherwise without the prior per-
mission of the copyright owners.

Printed in the United Kingdom by
Taylor & Francis (Printers) Ltd,
Rankine Road,
Basingstoke, Hampshire, RG24 0PR

Distributed in the United States of America and Canada
by International Publications Service Taylor & Francis Inc.,
114 East 32nd Street, New York, NY 10016.

British Library Cataloguing in Publication Data
Rohmert, Walter
   A new technique for job analysis.
   1. Job analysis
   I. Title II. Landau, Kurt
   III. Das arbeitswissenschaftliche Erhebungsverfahren
   zur Tätigkeitsanalyse (AET) *English*
   658. 3'06   HF5549.5.J6

ISBN 0 85066 251 6

Contents

		Page
Part I: Introduction to Job Analysis		7
1	What does "job analysis" mean?	9
2	What is the purpose of job analysis?	13
3	What is the AET?	15
4	How to carry out the AET applications	18
5	How to evaluate AET data	22
Part II: AET Manual		27
Introduction to the application of the classification codes		29
Part A – The Man-at-Work System		31
1	Work objects	33
1.1	Material work objects	33
1.1.1	Time allotment to work objects	33
1.1.2	Special material properties	34
1.1.3	Surface condition	34
1.1.4	Handling sensitivity	34
1.1.5	Shape	35
1.1.6	Size	35
1.1.7	Weight	36
1.1.8	Danger	36
1.2	Energy controlled in work activity	37
1.3	Information as a work object	37
1.4	People, animals, plants as work objects	37
2	Equipment	38
2.1	Work instruments	38
2.1.1	Work instruments modifying work materials	38
2.1.1.1	Work instruments for transforming the state of materials	38
2.1.1.2	Work instruments for transforming, transporting or storing energies or information	39
2.1.1.3	Work instruments for men, animals or living plants as work object	39
2.1.2	Work instruments for modifying the location of work materials	39
2.1.2.1	Stationary handling equipment	39
2.1.2.2	Movable handling equipment	39
2.1.3	Other work instruments	40
2.1.4	Separate consideration of control elements	40
2.1.4.1	Discrimination by the positioning organ	40
2.1.4.2	Discrimination by the number of positioning possibilities	40
2.2	Other equipment	41
2.2.1	Work instruments for state recognition	41
2.2.2	Technical auxiliaries aiding the human senses	41
2.2.3	Work seat, work table, work room	41

		Page
3	Working environment	44
3.1	Physical environment	44
3.1.1	Environmental influences	44
3.1.2	Work hazards or risk of occupational disease	48
3.2	Organizational and social environment	50
3.2.1	Organization of working time	50
3.2.2	Position of the job within the sequence of operations	53
3.2.3	Position of the job within the organizational hierarchy	54
3.2.4	Position of the activities within the communication system	58
3.3	Principles and methods of remuneration	60
3.3.1	Determination of the basis of remuneration	60
3.3.1.1	Legal basis of remuneration	60
3.3.1.2	Grading principles	60
3.3.1.3	Mode of payment	61
3.3.2	Determination of the remuneration method	61

Part B — Task Analysis 65

1	Tasks mainly related to material work objects	67
2	Tasks mainly related to information-processing and planning	69
3	Tasks mainly related to other persons	71
4	Number and frequency of task repetitions	72

Part C — Analysis of Work Demands 73

1	Field of work demands; reception of information	75
1.1	Dimensions of perception	75
1.1.1	Visual reception of information	75
1.1.2	Reception of information via sense of hearing	76
1.1.3	Reception of information via sense of touch or thermo-sensors of the skin	77
1.1.4	Reception of information via sense of smell or sense of taste	78
1.1.5	Proprioceptive information input	79
1.2	Modes of recognition	79
1.3	Accuracy of the reception of information	79
2	Field of work demands: decision	81
2.1	Complexity of decision	81
2.2	Urgency of decision	82
2.3	Required education	82
3	Field of job demands: action	86
3.1	Stress due to postural work (postures/body positions)	86
3.2	Stress due to static work	87
3.3	Stress due to heavy dynamic work	88
3.4	Stress due to active light work	88
3.5	Application of forces and frequency of motion	89

Part I: Introduction to Job Analysis

1 What does "job analysis" mean?

Legal regulations and the socio-political necessity of rendering working conditions more humane have promoted the process of analysing work tasks and requirements and the establishment of adequate guidelines for work design. Information about the workplace is a prerequisite in the fields of work-structuring and industrial safety, particularly as far as the humanization of working life is concerned. Traditionally, most workplace-related information is derived from work studies. These are done in order to create humane work sequences and working conditions, and at the same time maintain or even improve industrial profitability. *(Motivation for job analysis)*

Thus job analyses are basically nothing new. On the contrary: the development and application of work analysis methods can be traced back through several centuries. The fact that no less than 15 new methods have been published since 1970 is an indication of the many new uses of work analysis, but also of the shortcomings of earlier proceedings that have now been recognized. *(Development of job analysis)*

Many traditional methods of job analysis are oriented according to obsolete models of personality structures. Others are mainly based on energetical concepts that consider the physical and chemical environmental conditions. A general characteristic of these proceedures is their usefulness for a better evaluation of certain preferred forms of human work or work content, respectively.

The change that has occurred in work design, however, has above all resulted in a shift of work emphasis away from energetical work towards more informatory work (with increasing importance of sensorial, combinatorial and decision-related characteristics). This shift can hardly (if at all) be taken into account by the traditional processes (cf. Rohmert et al., 1975).

It is symptomatic that scientists, practitioners and representatives of the social parties do not use the idea and content of job analysis in a consistent way. First, the basic statements are usually not the same in the disciplines of ergonomics, industrial psychology and industrial sociology, and secondly, the methods of job analysis differ within any one discipline according to (cf. Frei, 1977). *(Basic concepts of job analysis)*

- the purpose of the analysis
- the object of the analysis
- the method of the analysis
- the underlying theoretical concept

as well as according to their degree of compliance to known criteria of quality such as validity, reliability etc.

In this article, we would like to describe job analysis as a subsection of a comprehensive system of work studies that covers the analysis of the individual components of man-at-work systems as well as the description. *(Definition of job analysis)*

and scaling of their interdependences. Job analysis starts from a model of work activity and assesses all relevant aspects of
- the work object
- the work resources
- the working environment
- the work tasks, and
- the work requirements

with regard to stress and strain considerations. Therefore, the basis of our understanding of job analysis is composed of the theoretical model of the man-at-work system and the concept of the simultaneous distinction and interdependence of stress and strain.

Model of the man-at-work system

Basically, the model of the man-at-work system has proved its value for the description and analysis of work (e. g. cf. REFA, 1978). Thus the idea of the system has been made available as a universal methodical instrument for the analysis of human work. In general terms, a system is a conceptually defined aggregate of individual objects or elements that are mutually related and possibly related to the environment, i. e. to everything outside the defined area (Kirchner, 1980).

Relation between man and work object

The model of the man-at-work system (Figure 1) is founded on the systematics of the basic types of work task (Rohmert, 1972). In each case, the man exerts an influence on a work object by means of methods. This influence leads to some feedback to the man. Both the work object and the man are situated in an environment defined by optionally selected limits. The environment is influenced by the relations between work object and man and in turn influences these relations. The relations between the man and the work object are employed systematically to obtain a work result; this involves the use of material, energy and information. The work result may act upon the man/work object relation and upon the environment in terms of quality and quantity.

Properties of the man-at-work system

Man-at-work systems are characterized as being purposeful, dynamic and concrete. The purpose and dynamics of the man-at-work system are derived from the object of work science, which is human work. Human work is always dynamic, i.e. continuously active, and aimed at a specific result. Human work is performed under certain internal and external conditions.

Conditions of human work

Internal conditions (e. g. muscular force, blood circulation, reaction time, manual skill, ability for team work) depend on the physical and psychological characteristics of the individual. External conditions are the working conditions (e. g. daily, weekly, monthly, yearly or lifetime working hours, length and distribution of rest allowances, situation of the workplace, type of working tool or machine, climate, dust, noise, mechanical vibrations). The way in which the worker is treated by colleagues or superiors (e. g. appreciation or criticism, approval or disapproval) also counts among these external conditions. However, it is not possible to consider all the external conditions mentioned here for job analysis. The characteristics, capabilities and skills of individuals can only be considered when they permit the description of the resulting work requirements in an objective, unambiguous and reconstructible manner. Even the mental attitude of a worker towards his work, the satisfaction that he gets out of the performance of his work and the treatment of the worker by his colleagues and superiors are not the subject of the job analysis concept described in this paper.

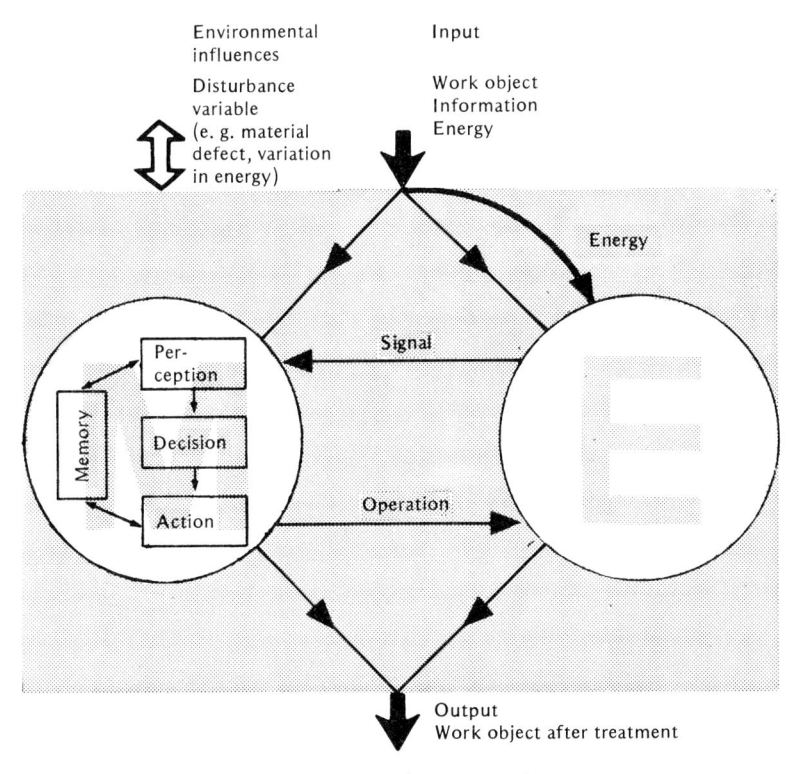

Figure 1: Model of the man-at-work system (REFA, 1971)

Nevertheless, this does not imply an isolated assessment of the man-at-work system, but rather its inclusion in the organizational, economic and social structure of the enterprise.

Besides the model of the man-at-work system, the stress/strain concept is another prerequisite for the evaluation of human work and thus also for job analysis (cf. Rohmert, 1972).

Stress/strain concept

The terminology of stress and strain has been derived from technical language and adapted to the human sphere. The performance of work through human activity and the use of technical means in the man-at-work system can be characterized by work tasks. From these can be derived the work requirements necessary for their fulfilment.

Stress intensity and stress duration

These are a measure of the energetical demands and the informatory difficulty of the work, which, together with the physical and organizational conditions, determine the stress to which the human being is subjected (Figure 2). Each stress quantity is composed of several partial stress quantities, which in turn may be subdivided into components of stress intensity and stress duration. Time intervals with constant stress intensities are called stress periods, i.e. the total stress duration is made up of a sequence of stress periods. The total amount of stress can be described by the intensity, the duration, the sequence, the overlapping and the timing of partial stress quantities within one shift (Laurig, 1980). The rather complex stress concept can thereby be subdivided into qualitative or quantitative dimensions, the determinants of which may be assessed by work analysis.

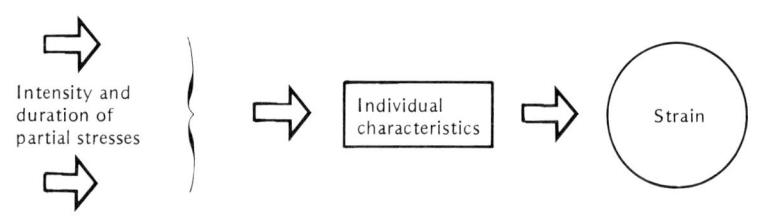

Figure 2: Stress/strain concept

Human characteristics, capabilities and skills

The stress determined by the work tasks and the working environment relates to human characteristics, capabilities and skills. Subjective strain in the worker results from stress as a function of the individual characteristics, capabilities and skills.

The criteria of objectiveness and repeatability exclude the assessment of the strain in work people through a job analysis procedure.

Nevertheless, strain can be an element of the discussion of stress. Generally, it can be concluded that the stress/strain concept furthers the breakdown of a rather complex total analysis of man-at-work systems into several subordinated dimensions that are easier to handle.

2 What is the purpose of job analysis?

As well as being based on the model of the man-at-work system and the stress/strain concept, the job analysis must be universal. This means that it must be applicable beyond the limits of individual enterprises, industries, or even countries. This is the only way to break away from the negative examples of the past, where the results of a job analysis case study applied only to the one specific case and were not transferable. The consequence of this was that the disciplines of ergonomics, industrial psychology, industrial medicine etc. gained little or no benefit from this kind of work study, because the results obtained were not generally applicable.

Universal job analysis

As seen by the practitioner, there are also aspects in favour of a comprehensive and universal solution. On the one hand, solutions that have been obtained in other areas and that are to be transferred can meaningfully be applied or adapted only if the circumstances of the situations in question are known and can be compared or correlated. On the other hand, ad hoc solutions that are merely partial are likely to miss the target of an optimum integral solution or even to create new problems in neighbouring areas. Through a good documentation of the case studies — which also implies adequate work studies — it is possible to accumulate a valuable set of experiences that can then be readily accessed and used in the solution of new problems (cf. Kirchner, 1982).

Advantages of universal job analysis

The concept of work analysis as a universal "wideband" method allows the solution of quite a variety of practical and scientific problems. Applications are possible in the fields of

Application of job analysis

1. Analysis of requirements and work design
   - Workplace documentation (e. g. with respect to the investigation of accidents and analysis of industrial safety)
   - Work design
   - Design of work resources
   - Systematic reduction of stress
2. Industrial organization
   - Preparation of organizational changes
   - Design of work sequences
   - Organizational work design and work-structuring
   - Organization of shift hours and rest allowances
   - Determination of rest allowances
3. Personnel management
   - Personnel recruitment
   - Personnel selection and placement
   - Integration of handicapped persons and other groups into the enterprise
   - Basic and advanced education
   - Work evaluation and remuneration

4. Vocational counselling and research
   - Health status in different job classes
   - Job classification with respect to requirements
   - Vocational counselling and information.

Better mutual adaptation of work and worker through job analysis

As seen by work scientists, the main purpose of applying job analysis is undoubtedly its usefulness in adapting the work to the worker by constructive and methodical work design, and in adapting the worker to his work by adequate education, exercise and selection. In particular, the systematic application of job analysis procedures can either completely replace extremely expensive physiological investigations or at least restrict them to small bottle-neck situations.

Types of job analysis user

Besides the investigation of possible uses of job analysis, the classification of its users is of particular interest. The analysis of the list of participants of work study seminars held by the authors of this paper yields the result that the participants come mainly from:

- Work study departments
- Organization and data processing
- Project groups concerned with work design
- Personnel management
- Industrial safety departments
- Members of the enterprises' medical services.

These users can be further classified by hierarchical position:

- Lower and medium management level (e. g. group managers, directors of work preparation departments)
- Staff members (e. g. medical service, ergonomics departments)
- Independent consultants and members of institutes.

3 What is the AET?

The origins of the AET (Arbeitswissenschaftliches Erhebungsverfahren zur Tätigkeitsanalyse; ergonomic job analysis procedure) date back to a study (Rohmert and Rutenfranz, 1975) ordered by the German government to investigate discrimination against women at work with respect to pay. A job analysis procedure was required that allowed a detailed investigation of work-load and strain within a given work system. At that time there was no job analysis procedure that could readily be used, although the PAQ (McCormick et al., 1969) seemed to provide a basis for the psychological items.

Origins of the AET

Starting from that particular research project, the AET (Landau et al., 1975; Landau, 1978; Rohmert and Landau, 1979) has been continuously developed over the last few years and applied in different industrial situations to analyse a wide range of shop-floor and management jobs (a total of about 4 000 job analyses).

The AET has been developed for the universal analysis of work systems, the content of work ranging from "production of forces" to "production of information". The procedures can be used for selection, placement, training, job classification, rehabilitation, job design, occupational medicine, work safety etc.

The AET is structured in three parts according to the theoretical concept described above: tasks, conditions of carrying out these tasks and the resulting demands upon the workperson (Figure 3).

AET structure

In part A (Work System Analysis) types and properties of work objects, the equipment to be used, and the physical, social and organizational working environment are documented on nominal and numerical scales. The behaviour requirements approach (Fleishman, 1975) is covered in part B (task analysis). On 31 numerically scaled items, tasks relating to material work objects and to abstract work objects and to man-related tasks are rated. Performing a task (as rated in part B) under the conditions documented in part A leads to job demands, which are evaluated in part C (job demand analyses) in three sections: perception, decision and response/activity.

Work system, tasks and demands at the workplaces are split up by the AET into a series of items, which are investigated by using the technique of observation/interviews carried out by trained analysts.

For economic reasons the total number of AET items cannot be complete in terms of the basic theoretical concept. Therefore, items have been selected that are significant for a greater number of work systems and permit differentiation between similar work systems.

Every AET item consists of an (underlined) question outlining the state of affairs to be grasped — under certain circumstances with examples as classification aids — and of the indication of a code for classifying this feature.

AET Items

Part A — Work System Analysis

1 Work objects
   1.1 material work objects (physical condition, special properties of the material, quality of surfaces, manipulation delicacy, form, size, weight, dangerousness)
   1.2 energy as work object
   1.3 information as work object
   1.4 man, animals, plants as work objects

2 Equipment
   2.1 working equipment
      2.1.1 equipment, tools, machinery to change the properties of work objects
      2.1.2 means of transport
      2.1.3 controls
   2.2 other equipment
      2.2.1 displays, measuring instruments
      2.2.2 technical aids to support human sense organs
      2.2.3 work chair, table, room

3 Work environment
   3.1 physical environment
      3.1.1 environmental influences
      3.1.2 dangerousness of work and risk of occupational diseases
   3.2 organizational and social environment
      3.2.1 temporal organization of work
      3.2.2 position in the organization of work sequence
      3.2.3 hierarchical position in the organization
      3.2.4 position in the communication system
   3.3 principles and methods of remuneration
      3.3.1 principles of remuneration
      3.3.2 methods of remuneration

Part B — Task Analysis

1 tasks relating to material work objects
2 tasks relating to abstract work objects
3 man-related tasks
4 number and repetitiveness of tasks

Part C — Job Demand Analysis

1 Demands on perception
   1.1 mode of perception
      1.1.1 visual
      1.1.2 auditory
      1.1.3 tactile
      1.1.4 olfactory
      1.1.5 proprioceptive
   1.2 absolute/relative evaluation of perceived information
   1.3 accuracy of perception

2 Demands for decision
   2.1 complexity of decision
   2.2 pressure of time
   2.3 required knowledge

3 Demands for response/activity
   3.1 body postures
   3.2 static work
   3.3 heavy muscular work
   3.4 light muscular work, active light work
   3.5 strenuousness and frequency of movements

Figure 3: Contents of the AET

Item-rating can only be done by means of the corresponding code. The code of significance (S) and exclusive code (E) are used for indicating the level of stress, the code "amount of time" for the duration (D), the code of frequency (F) for characterizing the temporal distribution and position of stress sections, and alternative and nominal scale special codes for documenting the working system characteristics (cf. McCormick et al., 1969). The codes and their use are described in detail on page 29. All AET items must be classified in order to answer all questions. The items of the job demand analysis — which could possibly be difficult to answer for a practician who is not sufficiently trained in ergonomics — contain additional classification aids in the form of "task or activity scales". The activity scale, based on previously investigated data, contains a series of grades of mainly illustrative tasks and activities. Equivalent to the ascending rating scale we may suppose — at least approximately — an intensified strain.

4 How to carry out the AET applications

Selection and training of AET analysts

The successful application of the AET depends particularly on a thorough knowledge ot the structure and content of the procedure, its practical use in the field, the application of statistical evaluation techniques and the selection and training of the AET analysts. Procurement of the necessary AET material is followed by the selection and training of AET analysts. If the raters are work analysts who are acquainted with ergonomic procedures, an AET training using a manual or the programmed instruction (in preparation) is sufficient. If the analysis is to be performed by work and time-study specialists, supervisors or workers themselves, participation in an AET training seminar is recommended in addition to training with the AET manual/programmed instruction. The authors conduct special REFA seminars for this purpose.[1]

Before applying the AET for the first time the job analyst should thoroughly study the structure and handling of the questionnaire. This includes — besides the reading of the instructions for use — the study of the individual sections and items, since only thorough understanding of the questionnaire and knowledge of the content of the job elements enable an objective description and evaluation of a workplace.

As a general principle it must be observed that the evaluation as a part of a job analysis is related to job and workplace; that means that personal characteristics of the worker (e. g. dissatisfaction, overstrain) cannot be taken into consideration in the AET analysis.

Course of a job analysis

The course of a job analysis with the AET is shown schematically in the following survey (Figure 4); the most important details will be discussed there.

Preparatory to a job analysis in an enterprise, a preliminary discussion with the management is necessary in most cases, e. g. about purpose and date of investigation, or selection of the workplace. Moreover, the works council or personnel council should be informed. Afterwards, the incumbent and his direct superior should be informed about the form, purpose and course of the analysis.

Observation/ interview

The analysis of a job is done in the form of an observation/interview, which means that the necessary analytical data are collected first by the observation of a job and working environment, and secondly by interviewing the incumbent and his supervisor. Priority must be given to the analyst's observation. The interview, however, is used to ascertain in addition those job characteristics that could not be determined by observation.

The course of the observation/interview can be varied according to the different work forms and the repetitiveness of the work process of the jobs in question. In the case of a mainly physical activity all important job characteristics, including work process, work objects, working tools and

---

1) Please contact: Prof. Dr.-Ing. Landau, Berliner Straße 23, D-6103 Griesheim.

```
                    ┌─────────────────────────┐
                    │ preliminary discussion  │
                    │ with managers and works │
                    │ council or personnel    │
                    │ council                 │
                    └─────────────────────────┘
                                │
                    ┌─────────────────────────┐
                    │ information of the      │
                    │ worker and the direct   │
                    │ supervisor              │
                    └─────────────────────────┘
                                │
                         ( work form )
```

work form mainly physical | work form mainly non-physical

observation at the workplace
e.g. operation
work objects, working agents, work chair, table, room
working environment
contacts with colleagues and superiors

interview with job incumbent,
e.g. job designation
job description
work organization
responsibility
authority to give instructions

(degree of repetitiveness of operations)

observation at the workplace
e.g. work chair, table, room
working environment
contacts with colleagues and supervisors

high | low

detailed observation
e.g. operation
work object, working agents, exposure to danger

talk with supervisor or job incumbent
e.g. variation of work object and working agent, lot size

talk with supervisor in order to control and supplement already investigated data

e.g. special tasks and requirements
work organization
responsibility
authority to give instructions
task allocation
exposure to danger
occupational diseases
remuneration

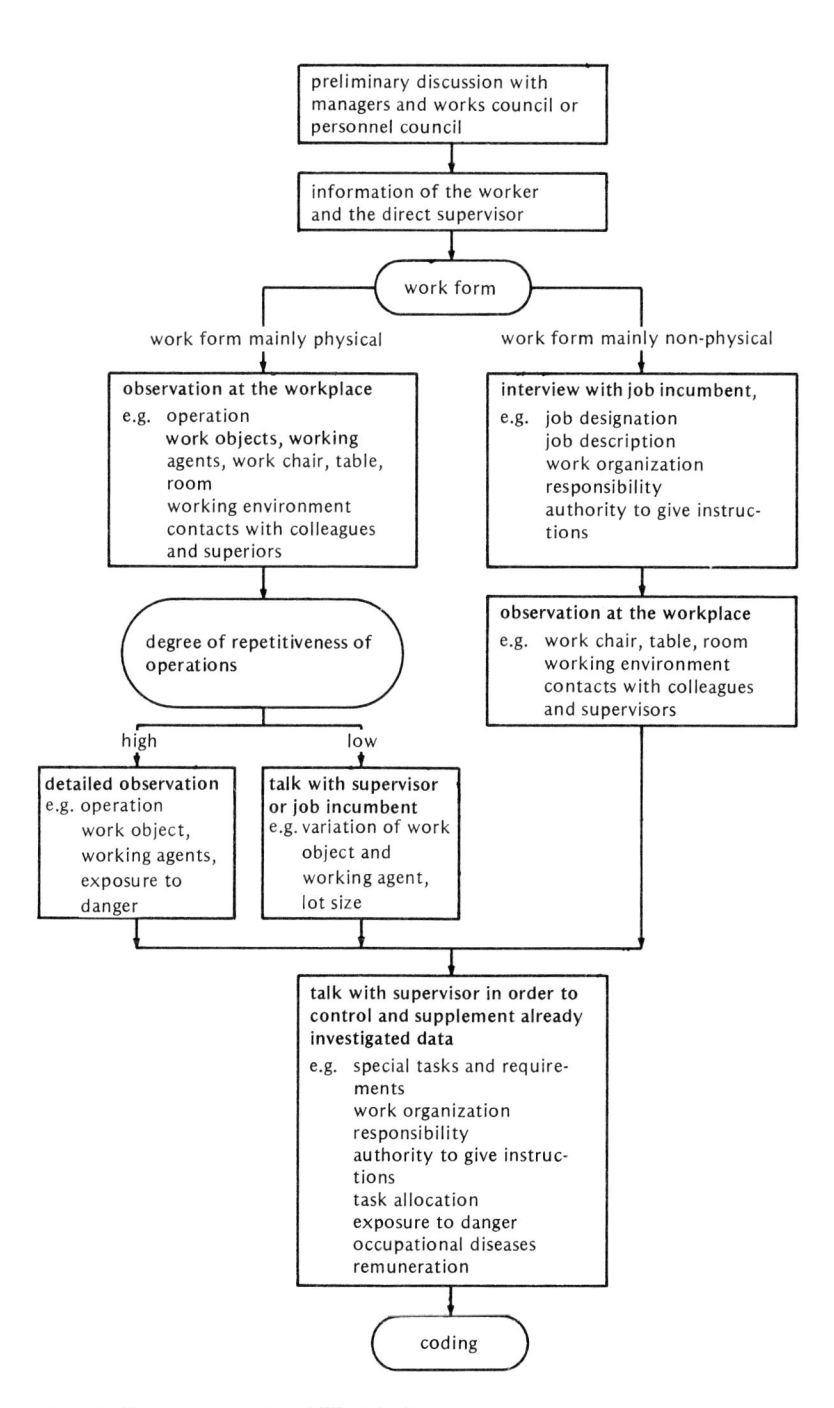

(coding)

Figure 4: How to carry out an AET analysis

environment, could be determined in the first place by a fairly long observation at the workplace. This applies especially to the case of high repetitiveness. In the case of a low degree of repetition a supplementary talk with the supervisor or incumbent — e. g. about kind and frequency of variation of work object and working agents, lot size etc. — could become necessary. In the case of a mainly non-physical activity it is advisable to carry out — previous to the observation — an interview at the workplace — e. g. about tasks, organization, responsibility, authority to give instructions — because these activities mostly show a low repetitiveness; besides, mere observation is insufficient for the understanding and evaluation of the work content.

In conclusion, a talk with the supervisor aiming at a check and completion of previously collected data and a settling of questions not yet answered — e. g. about work organization, task allocation, remuneration — is required in most cases.

The necessary time for an observation/interview amounts to approx. 2 to 3 hours, depending on the analyst's practical routine, the type of job and the repetition rate of work processes.

Coding of AET items

The AET analysis is directly followed by the coding of the AET characteristics. This coding has to be done on marking documents as shown in Figure 5.

After having performed several analyses, AET-users send their coding forms to a computer centre, which evaluates the results and guarantees the protection of the user's data.[1]

---

1) Please contact: Verlag Hans Huber, Länggass-Straße 76, CH-3000 Bern 9.

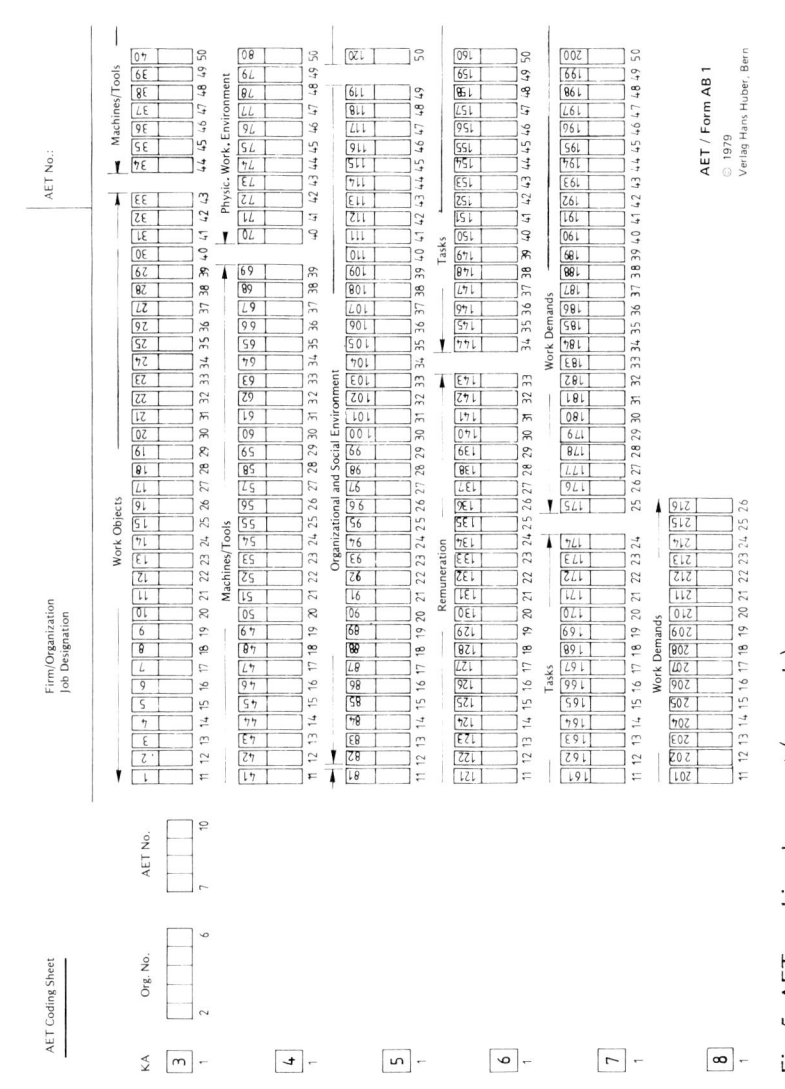

Fig. 5: AET marking document (example)

5 How to evaluate AET data

Further information on evaluation procedures and results of AET applications are shown in the AET case-study book by Landau and Rohmert (1981).

Analysis of frequency distributions

The following representations (Figures 6 and 7) explain how firm-, branch- and sex-related position groups that were analysed by means of job analysis procedures can be evaluated in a simple way by analysing the frequency distributions of levels. Figure 6 shows the percentages of the feature levels "high" and "extreme" for selected items of the analysis procedure. It can be mentioned as an example that at 7 % of the workplaces of male workers the climatic stress is rated "high" or "extreme". In the field of pre-assembly, 13 % of the workplaces of male workers show a high or extreme climatic stress. In the assembly department, an extraordinary climatic stress is to be found. Effects of mechanical vibrations in the fabrication department are rated "high" or "extreme" at 87 % of the workplaces occupied by men and at 28 % of the workplaces occupied by women. This discussion can be continued in the same way for all characteristics of the procedure. If large-scale data collections related to branches are available, it is possible to carry out an analysis of frequency distributions of levels of job analysis data for individual branches. This is shown in Figure 7.

AET-profiles

A representation of the scores obtained by the groups of characteristics in the form of profiles is suitable to give a graphic survey of the extent or the duration of stress during execution of given activities or groups of activities. This type of evaluation of the data derived from job analysis is designated "profile analysis", regardless of the fact that this term may possibly be used in a different sense by other disciplines (Figure 8). The classification of the characteristics that is necessary for the execution of a profile analysis can be derived from the structural items of the job analysis procedure as is done for the AET, or a job profile can be established on the basis of a factor analysis of the data derived from job analysis. This method was used by McCormick, Mecham and Jeanneret (1972).

Both classification and summing-up of the characteristics according to structural criteria of the analysis procedure and according to the results of a factor analysis require metric scaling of data. A transformation of data derived from job analysis to interval or proportional scales always involves the danger of adding wrong information. For calculation of the profile heights, the scores obtained for the characteristics of an activity or group of activities can be related in a first approximation to the maximum possible scores of the groups of characteristics. This calculation following the "mediation principle" is determined by the number of the job analysis characteristics assigned to each fact. This implies that there is no comparability between the profile columns of the same profile analysis. The comparison of the profile heights of different jobs, however, is possible and even intended.

Factor analysis

Methods of *factor analysis* are suitable for the determination of stress factors that are characteristic for a job but cannot be measured or observed directly. In the process, extensive data being assessed are normally reduced to a small number of independent hypothetical dimensions or factors.

AET item	fabrication male	fabrication female	% level 4 + 5 pre-assembly male	pre-assembly female	assembly male	assembly female
⋮ climate (71) ⋮	7	7	13	1	0	0
⋮ vibrations (78) ⋮	87	25	2	0	18	17
⋮ accuracy of perception (17) ⋮	61	57	0	0	0	0
⋮ decision-making under pressure of time (19) ⋮	0	0	39	39	91	89

Figure 6: Frequency distributions of levels for in-company applications

AET item	occupied by women	occupied by men	% level 4 + 5 metal working ind.	chemistry
active light work	32	6	24	22
vis. ident. of surface structures	34	25	34	39
combining	2	20	3	0
⋮ noise ⋮	95	40	35	43
responsibility ⋮	1	51	22	21

Example:
Sex- and branch-related list of the classifications "high" and "extreme" for selected AET characteristics

Figure 7: Frequency distribution of levels for AET characteristics

A series of publications is available concerning the application of factor analysis procedures to data gathered by job analysis (cf. Jeanneret, 1969; Frieling and Hoyos, 1978; Luczak et al., 1977; McCormick, Jeanneret and Mecham, 1972) and it would be too much to deal with this problem at this point.

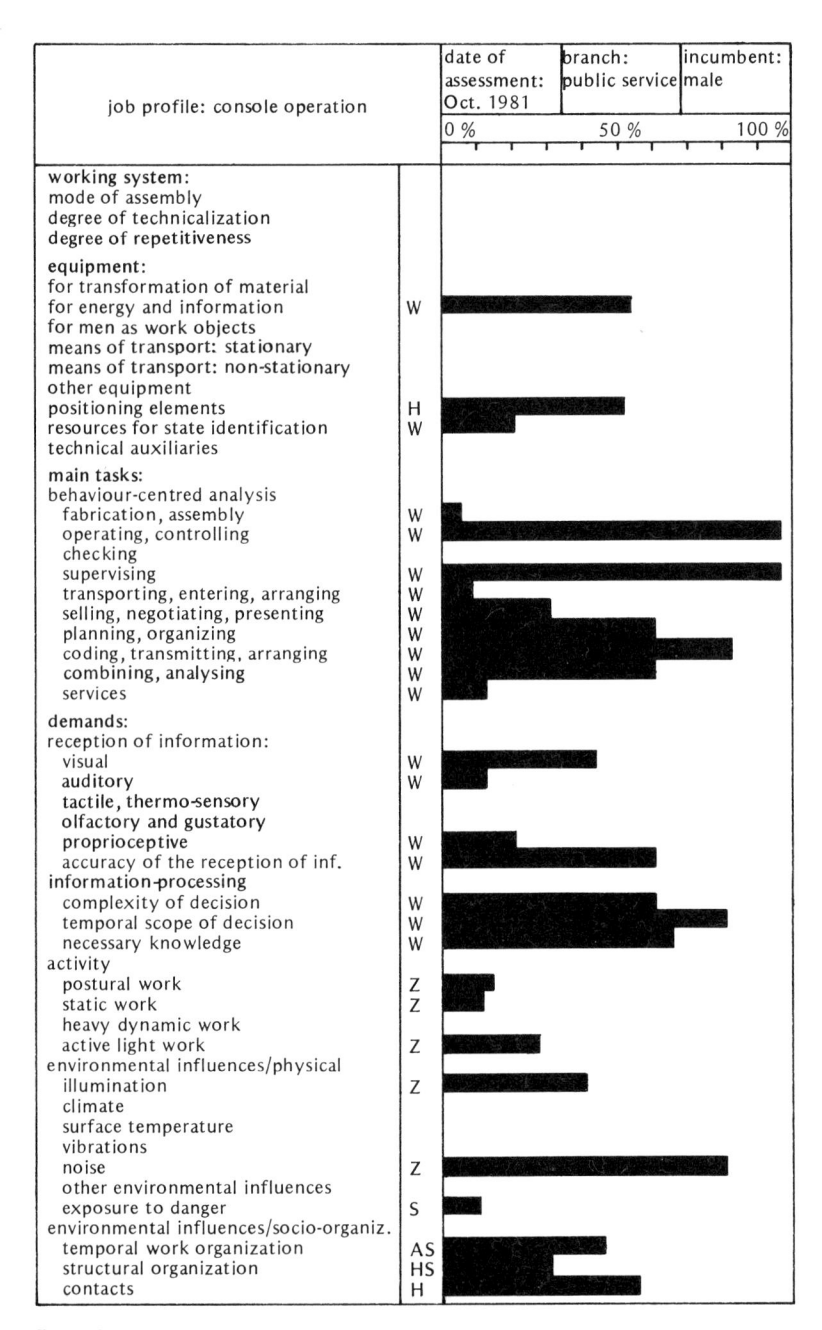

job profile: console operation		date of assessment: Oct. 1981	branch: public service	incumbent: male
		0 %	50 %	100 %

working system:		
mode of assembly		
degree of technicalization		
degree of repetitiveness		
equipment:		
for transformation of material		
for energy and information	W	
for men as work objects		
means of transport: stationary		
means of transport: non-stationary		
other equipment		
positioning elements	H	
resources for state identification	W	
technical auxiliaries		
main tasks:		
behaviour-centred analysis		
fabrication, assembly	W	
operating, controlling	W	
checking		
supervising	W	
transporting, entering, arranging	W	
selling, negotiating, presenting	W	
planning, organizing	W	
coding, transmitting, arranging	W	
combining, analysing	W	
services	W	
demands:		
reception of information:		
visual	W	
auditory	W	
tactile, thermo-sensory		
olfactory and gustatory		
proprioceptive	W	
accuracy of the reception of inf.	W	
information-processing		
complexity of decision	W	
temporal scope of decision	W	
necessary knowledge	W	
activity		
postural work	Z	
static work	Z	
heavy dynamic work		
active light work	Z	
environmental influences/physical		
illumination	Z	
climate		
surface temperature		
vibrations		
noise	Z	
other environmental influences		
exposure to danger	S	
environmental influences/socio-organiz.		
temporal work organization	AS	
structural organization	HS	
contacts	H	

Figure 8: AET job profile: console operation

The application of *cluster-analytic methods* to job analysis data aims mainly at a reduction of the quantity of data. Apart from the generation of job groups, numerical relationships between the job groups must be established and interpreted. When carrying out the classification of activities one must examine to what extent a descriptive method turns into a explicative one, i. e. to what extent conclusions can be drawn from the numerical relationship for the elaboration of a taxonomy of human activities. As far as ergonomics is concerned, the effects of specific work content on group composition and on numerical relationships between the groups are of particular interest.With the breakdown of large ergonomic data pools into smaller and comprehensive groups arises the possibility of further investigating groupscores, i. e. the positions that are most representative for the group as well as the respective workers, using (rather expensive) methods of occupational physiology and psychology. The results of these further detailed investigations can subsequently be re-transferred to the remaining positions in the same job group that are similar as far as strain is concerned, observing the applicable limits.

Figure 9 shows one application of cluster analysis methods on AET data.

A hierarchic cluster analysis has been carried out for 24 buying, selling or trading jobs and represented in the form of an activity dendrogram. In the analysis, two groups clearly separated with respect to working posture can be distinguished.

1. Prevailing working posture is normal standing with a small quota of the postures "bent standing", "crouching" and "kneeling". This group covers the working postures typical of selling activities in department stores and retail shops.
2. Prevailing body posture is sitting with a small quota of standing. This posture is typical of commercial activities not to be classed with the retail trade. Significant activities in this group range in the field of marketing consulation and sales promotion.

To sum up it can be said that the use of statistical data analysis procedures for AET data provides significant aid in interpretation and data reduction. It is the cluster analysis in particular that allows an economical application of job analysis in the fields of work design, demand analysis, personnel management, work and occupational research, detection of accident causes etc.

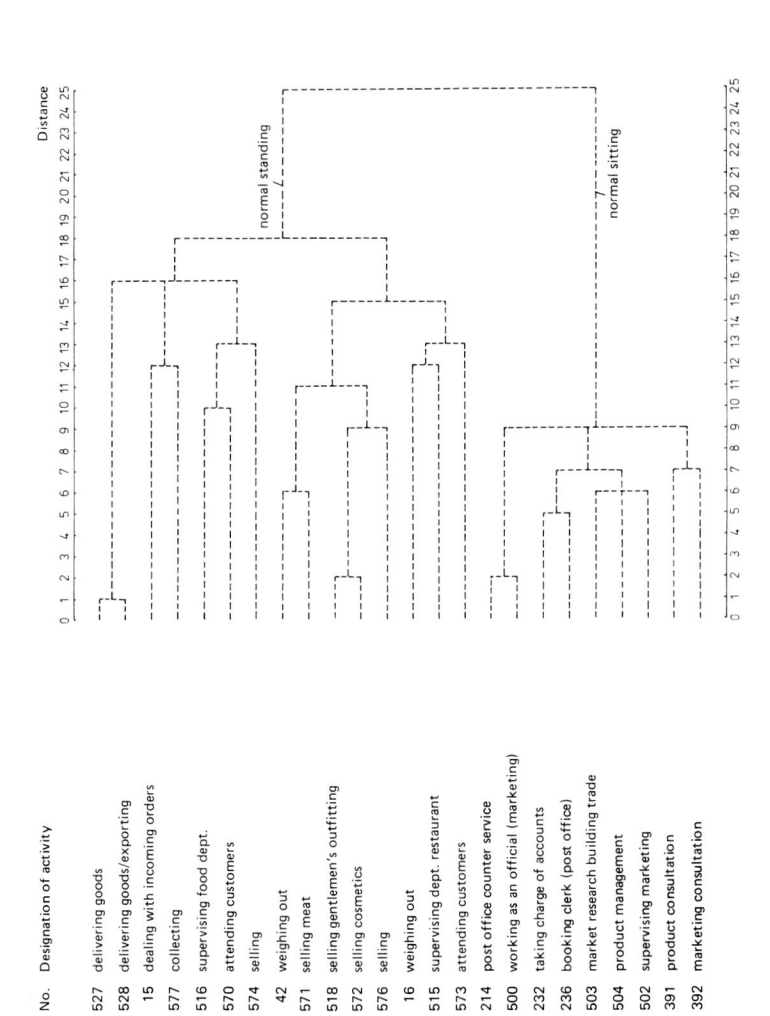

No.	Designation of activity
527	delivering goods
528	delivering goods/exporting
15	dealing with incoming orders
577	collecting
516	supervising food dept.
570	attending customers
574	selling
42	weighing out
571	selling meat
518	selling gentlemen's outfitting
572	selling cosmetics
576	selling
16	weighing out
515	supervising dept. restaurant
573	attending customers
214	post office counter service
500	working as an official (marketing)
232	taking charge of accounts
236	booking clerk (post office)
503	market research building trade
504	product management
502	supervising marketing
391	product consultation
392	marketing consultation

Figur 9: AET cluster analysis (example)

Part II: AET Manual

Introduction to the application of the classification codes

The job characteristics that are analysed with the AET are subsequently evaluated by means of standardized classification codes designed to obtain comparable job information in a relatively short period of time.

Multi-level codes are used to evaluate the extent of stress — significance (S) and exclusive (E) codes — and the duration of stress — duration (D) and frequency (F).

Additional single- and multi-level codes are used on the nominal scale level to classify the state of job design.

The standard codes S, D, F and A are not referenced separately in the text but are contained in the appendix "Standard Codes".

The exclusive (E) code stands with each characteristic for the classification for which it is to be used.

Note: When classifying a characteristic only use the corresponding code (= capital letter preceding the individual characteristic).

Significance (S) code

When using the significance code for classification of a work element, the importance of an aspect for the work output should be estimated in relation to the other aspects.

It would therefore be inadmissible when analysing the activity "motor-car-driving" to rate the turning of the ignition key with level 5 (= very high importance) merely because without this the whole activity would not take place. On the contrary, this non-recurring operation should be considered in relation to other more frequently occurring operations, such as operating the clutch, applying the brakes etc.

Exclusive (E) code

The exclusive code is always related to only one specific question. It enables classification of a factor that does not fit under the standard classification code. There are two variants of this code:

a) *the classifying code* (nominal scale level)
    Multiple properties, e. g. of a working instrument, are ascertained. Normally, only one of these properties can apply. If there are in the opinion of the analyst several important properties, the one must be selected that is most significant for the incumbent's work;
b) *the strain-related code* (numerical scale level)
    This code is used to describe a strain associated with stress exerted on the working person. The code is graded in steps $0 - 5$, which indicates increasing strain.

Duration (D) code

The duration code is based on a shift lasting 8 hours. For cases where it is used as a code, an 8-hour shift is always assumed even if the incumbent is a part-time worker. This method allows the possibility of comparing the work content of jobs with different shift hours.

The gaps between the code levels are deliberately unequal because it is assumed that an analyst may be able to classify the lower and upper range fairly precisely, while it is more difficult to make such clear distinctions in the large medium range (between 1/3 and 2/3 of shift time).

Frequency (F) code

The frequency code is used to relate the frequency with which a work element occurs to the frequency of other factors in the job to be described.

Alternative (A) code

The alternative code asks merely about the presence of a specific factor. Does the work characteristic in question apply?

Standard codes to be used

Code: *Does the work characteristic apply or not? (A)*
- 0 Does not apply
- 1 Does apply

Code: *Frequency (F)*
- 0 Does not apply
- 1 Very infrequent
- 2 Infrequent
- 3 Average
- 4 Frequent
- 5 Very frequent

Code: *Significance for the work (S)*
- 0 Does not apply
- 2 Very minor
- 2 Minor
- 3 Average
- 4 High
- 5 Very high

The Duration (D) code should be related to an (assumed) shift of 8 hours.

Code: *Duration (D)*
- 0 Does not apply (or is *very* infrequent)
- 1 Less than 10 % (50 min) of shift time
- 2 Less than 30 % (160 min) of shift time
- 3 Between 30 % (160 min) and 60 % (320 min) of shift time
- 4 More than 60 % (320 min) of shift time
- 5 Almost continually during whole shift

Part A: The Man-at-Work System

1 Work objects

> Work objects are all materials, goods, energies and information that are transformed in the man-at-work system in conformity with the work task.

The duties of the incumbent may include tasks to be performed on several work objects. The duration (D) code classifies the individual work objects (which differ in type and properties) on which the working person exercises influence during his work.

1.1 Material work objects

CNO*	CC*	Description of characteristics
1	D	During what proportion of the *time* does the incumbent perform tasks involving these *work objects*?

For example, workpieces, packing material, fruit, meat, chemicals. These types of object are normally encountered with mainly physical work forms (muscular and sensory-motor work, e. g. transportation activities, assembling, crane-driving).

If this characteristic is rated D = O, continue with characteristic 32.

Material objects are further subdivided into the following.

1.1.1 Time allotment to work objects

2	D	During what proportion of the *time* does the incumbent perform tasks involving *solid work materials?*

applies, for example, to:
turning or locksmith's work, assembling, construction activities, agricultural and forestry work, stone-working.

3	D	During what proportion of the *time* does the incumbent perform tasks involving *liquids?*

applies, for example, to:
chemistry, production of beverages, painting and varnishing, moulding

4	D	During what proportion of the *time* does the incumbent perform tasks involving the use of *gases?*

extraction of natural gas, evaporator control, generation of welding gas

---
*) CNO: No. of characteristic
*) CC: Characteristic code

CNO	CC	Description of characteristics

1.1.2 Special material properties

Indicate the proportion of *time* during which the incumbent performs tasks involving work materials with *special properties.*

5	D	Work material *brittle* (risk of breakage)
		glasses, stoneware, pastry (biscuits), carbide tools
6	D	Work material *elastic/flexible*
		helical springs, leaf springs, linoleum, roofing paper
7	D	Work material *plastic*
		doughs, glues, cements, waxes, gelatine

1.1.3 Surface condition

Answer questions 8 – 10 indicating the individual proportion of *time* during which the incumbent manipulates work materials with *extreme surface conditions.*

8	D	Work material extremely *dirty*
		soot, waste water, coal, ash, faecal matters, waste
9	D	Work material extremely *greasy/sticky/slippery*
		glues, technical oils and greases, adhesives, waxes, soaps
10	D	Work material extremely *sharp-edged* (cutting or rough)
		constructional glass, non-deburred metal workpieces, abrasives, bricks, building lumber

1.1.4 Handling sensitivity

Answer questions 11 – 13 indicating the individual proportions of *time* during which the incumbent controls or works on materials that require extreme *sensitivity* in *handling.*

11	D	Work material *scratch-sensitive*
		e. g. jewellery, polished surfaces, films, acrylic glass
12	D	Work material *easily deformed or folded*
		e. g. linoleum, plaster building panels, wire, car body sheet, candles, cakes and tarts
13	D	Work material *brittle or easily torn*
		e. g. paper webs, insulation material for building purposes.

CNO	CC	Description of characteristics

1.1.5 Shape

Answer questions 14 – 18 indicating the proportion of *time* during which the incumbent performs tasks on work materials that have *special shapes*.

14	D	Work with *granules*
		e. g. grain, sand, pellets, washing detergents, artificial fertilizers
15	D	Work with *tangled materials*
		e. g. wire, screws, loose textiles, foamed plastics, loose yarn
16	D	Work with *sheet-like objects*
		e. g. paper webs, webs of cloth, synthetic foils, adhesive ribbons, roofing papers
17	D	Work with *flat parts or with blocks*
		e. g. punched parts, gears, wall panels, engine blocks, pallets, box-shaped containers
18	D	Work with *bars or rolls*
		e. g. axles, shafts, bottles, tubes, wound-up tangled material

The sum of the proportion of time allotted to characteristics 14 – 18 must be equal to the duration of the incumbent's work with material objects, characteristic 2.

1.1.6 Size

Answer questions 19 – 21 indicating the individual proportions of *time* during which the incumbent performs tasks on work materials of *different sizes*.

19	D	Work materials *very small/small*
		e. g. components used in watch-making, transistors, screws, gears for precision mechanical purposes, electric bulbs (objects can be manipulated with fingers)
20	D	Work materials *medium-sized*
		e. g. car wheel-rims, electric domestic appliances (objects can be manipulated with hands)
21	D	Work materials *large/very large*
		e. g. machine foundations, propulsion units, pieces of furniture (objects can be manipulated by means of handling equipment or with hands, arms or whole body)

CNO	CC	Description of characteristics

1.1.7 Weight

Answer questions 22 − 24 indicating the individual proportions of *time* during which the incumbent performs tasks involving work materials of *different weights*.

22	D	*Low* weight

    objects weighing up to 1 kg, can normally be manipulated with fingers or hands

23	D	*Medium* weight

    1 − 10 kg, can normally be manipulated with hands

24		*Heavy* weight

    more than 10 kg, can partly be manipulated by one person without using additional auxiliaries, partly including the use of handling equipment and hoisting machines

1.1.8 Danger

Answer questions 25 − 30 indicating the individual proportions of *time* during which the incumbent performs tasks involving *dangerous work materials*.

25	D	Work materials that are *explosive*

    e. g. explosives and igniting mixtures, ammunition, fireworks

26	D	Work materials that are *conducive to fire or inflammable*

    e. g. petrol, technical oils, lacquers and varnishes

27	D	Work materials that are *poisonous or caustic*

    e. g. basic chemicals, chemical-technical materials, plant protectives, cleaning materials

28	D	Work materials that are *radioactive*

    e. g. uranium concentrate, nuclear materials

29	D	Work materials *irritating skin or mucous membrane*

    e. g. quartz, asbestos, Thomas meal, flax, raw cotton

30	D	Work materials *causing other health hazards*

    If characteristic 1 is rated D = 5, continue with characteristic 34.

| CNO | CC | Description of characteristics |

1.2 Energy controlled in work activity

| 31 | D | Indicate the proportion of *time* during which the control of *energy* is the most important work requirement for the incumbent.

The incumbent does not generate the energy. Instead, he or she controls the use of energy via a control console for the positioning of parts.

If the generation, transport and storage of, e. g., electrical energy or thermal energy are of particular significance for the incumbent's work, then the corresponding duration code level must be indicated.

This characteristic does not apply to the use of energy-powered auxiliaries (such as fork-lifts)!

If characteristic 31 is rated D = 5, continue with characteristic 34.

1.3 Information as a work object

Information should be graded as a work object in a man-at-work system where the material work objects are less significant for the incumbent's work than *abstract work objects*.

| 32 | D | Indicate the proportion of *time* during which the work of the incumbent involves mental skills.

(Organizing, designing, acquisitioning, forming sales strategies, examining the traffic situation.) The information as a work object has particular significance in man-at-work systems including mainly non-physical activities (e. g. administrative work, translating, dictating, designing).

If characteristic 32 is rated D = 5, continue with characteristic 34.

1.4 People, animals, plants as work objects

| 33 | D | Indicate the proportion of *time* during which the incumbent deals primarily with *human beings, animals or living plants* as the most important work object.

When analysing the work of physicians, nurses, psychologists, teachers etc., *human beings* (sick people, the handicapped, babies) are to be evaluated as work *objects*.

Do not evaluate contact with special categories of persons, such as the handicapped or the elderly, which is involved to a certain extent in the work of every incumbent.

Animals or living plants are work objects for agricultural and horticultural activities, for the keeping of experimental animals, for running zoological gardens etc.

2 Equipment

This term comprises all devices or machinery used to carry out a performed task in a man-at-work system.

Of these, *work instruments* are devices that perform technical work by utilizing physical, chemical or biological principles.

2.1 Work instruments

2.1.1 Work instruments modifying work materials

CNO	CC	Description of characteristics

2.1.1.1 Work instruments for transforming the state of materials

34	S	Use of *stationary* work instruments for material production, shaping and reshaping (non-cutting transformation) e. g. blast furnace plants, chemical processing plants, moulding machines in foundries, deep-drawing presses
35	S	Use of *stationary* work instruments for separating, abrading, cleaning and chip-cutting e. g. lathes, milling, sawing and planing machines, cleaning and washing plants
36	S	Use of *stationary* work instruments for combining, joining, coating and material-transforming e. g. folding, stapling and riveting machines, sewing machines, spraying and dipping plants; work instruments for hardening, nitriding, chromizing; filling and cooling plants; also quasi-stationary work instruments such as a pneumatic screw-driver
37	S	Use of work instruments manually operated by incumbents for the transformation of materials e. g. drilling machines, lawn-mowers
38	S	Use of *power-driven* work instruments for material transformation e. g. harvester-threshers, road construction machines Only classify *work instruments;* handling equipment is to be classified under characteristics 46 – 50.
39	S	Use of *movable* work instruments for shaping or reshaping e. g. hammer

CNO	CC	Description of characteristics
40	S	Use of *movable* work instruments for separating, abrading, cleaning and cutting e. g. saw, scissors, chisel, abrasive paper
41	S	Use of *movable* work instruments for combining and coating e. g. screwdriver, wrench, brush, paint roller, soldering-iron (Office supplies are to be classified under characteristic 44.)

2.1.1.2 Work instruments for transforming, transporting or storing energies or information

42	S	Use of work instruments for *energy transformation* e. g. transformers, heat exchangers, heating and cooling plants
43	S	Use of work instruments for *transforming information* e. g. electronic data processing (EDP) systems, word processors, typewriters, calculators, accounting and billing machines
44	S	Work instruments used for the purpose of *organizing* office supplies, such as filing cards, EDP printouts, disposition boards, writing materials (pencils, fountain pens etc.); telephone

Under characteristic 44 avoid including complex energy-powered devices, which are classified under characteristic 43.

2.1.1.3 Work instruments for men, animals or living plants as work object

45	S	Use of stationary or movable work instruments for human beings, animals or living plants as work objects e. g. dentist's drill, scissors, radiation treatment device

2.1.2 Work instruments for modifying the location of work materials

2.1.2.1 Stationary handling equipment

46	S	Use of *continuous conveyors* e. g. suspension conveyors
47	S	Use of *hoisting and lifting machines* e. g. pulley blocks, travelling cranes, passenger elevators, lifting platforms

2.1.2.2 Movable handling equipment

48	S	Use of *manually operated* movable handling equipment e. g. wheelbarrow

CNO	CC	Description of characteristics
49	S	Conducting and operating *motorized* transport facilities e. g. motor cars, lorries, buses, trains, freighters, aeroplanes
50	S	Conducting and operating *other motorized mobile facilities* e. g. vehicles not primarily used in public traffic: electric trucks, fork-lift trucks, self-propelled lawn-mowers

2.1.3 Other work instruments

51	S	Use of other work instruments e. g. jigs and fixtures, tool mountings

2.1.4 Separate consideration of control elements

The following separate classification of control elements is subdivided into: (1) the *positioning organ;* and (2) the *positioning possibilities*. Each control element must be classified *twice*.

Read through all questions in paragraph 2.1.4.1 *before* starting the classification!

2.1.4.1 Discrimination by the positioning organ

52	F	Use of control elements actuated with *fingers* e. g. adjusting knobs, ignition keys, keyboard, wire-releasing devices, thumb slides, toggle and rocking switches, dials
53	F	Use of *manually actuated* controls actuation by hand (one or both) or by arm, e. g. operating lever, rudder, handwheel, crank
54	F	Use of *foot-operated* controls actuation by foot or by leg, e. g. brake, clutch and accelerator pedals; toggle lever

2.1.4.2 Discrimination by the number of positioning possibilities

55	F	Actuation of *service switches* (one or two positioning possibilities) Control elements with two possible positions are actuated to regulate the flow of energy, information or materials, or to adjust the position of component parts, e. g. light switch, ignition switch. Note that keyboards are to be classified under characteristic 57.

CNO	CC	Description of characteristics
56	F	Actuation of *control elements with three or more positioning possibilities* e. g. gear-shift mechanism, programme dials
57	F	Use of *keyboards* The incumbent operates, e. g., a typewriter or calculator, an accounting machine, an EDP console or piano keyboard by using one or several fingers.
58	F	Use of *control elements for continuous (stepless) adjustments* controls for switching-on and stepless regulation of work instruments, e. g. volume control on radio receiver, room thermostat, stepless regulation of the speed of a drilling machine Note that the use of micrometer screws and slide gauges is to be classified under characteristic 59.

2.2 Other equipment

2.2.1 Work instruments for state recognition

59	S	Use of *measuring devices* standard and precision measuring instruments for state recognition: ruler, folding rule, thermometer, water-level, spring balance, micrometer screw, stop-watch etc.
60	S	Use of *other instruments for state recognition* e. g. radar sets, X-ray units, TV monitors, telescopes, mirrors; functional, warning and monitoring indicators

2.2.2 Technical auxiliaries aiding the human senses

61	S	Use of technical auxiliaries supporting the senses e. g. magnifiers, sound amplifiers, auxiliary devices for handicapped workpeople

2.2.3 Work seat, work table, work room

62	E	*Work seat* Classify the work seat predominantly used by the incumbent. 0 No work seat 1 Work seat fully adjustable (adjustable seat back, depth and height of seat) 2 Work seat not adjustable 3 Stool 4 Standing seat

CNO	CC	Description of characteristics
63	E	*Foot rest* 0 No foot rest 1 Adjustable foot rest 2 Non-adjustable foot rest
64	E	*Work table* 0 No work table/work bench 1 Work table adjustable in height and angular position 2 Non-adjustable work table/work bench 3 Standing desk/console
65	E	*Arm rests* (free arm rests, arm rests mounted to seats and to work instruments or equipment) 0 No arm rest 1 Adjustable arm rest 2 Non-adjustable arm rest
66	E	*Number of persons working in the office space* How many workplaces exist in the office space where the incumbent works most of the time? 0 Does not apply 1 Single workplace 2 Workplaces for 2 persons 3 Workplaces for 3 − 6 persons 4 Workplaces for 7 − 30 persons 5 Workplaces for more than 30 persons
67	E	*Number of persons working in workshop or depot* How many workplaces exist in the workshop or the room in which the incumbent works most of the time? 0 Does not apply 1 Single workplace 2 Workplaces for 2 persons 3 Workplaces for 3 − 6 persons 4 Workplaces for 7 − 30 persons 5 Workplaces for more than 30 persons
68	E	*Size of work room (m^2)* Use the code below to indicate the approximate size of the work room where the incumbent works most of the time.

CNO	CC	Description of characteristics

0 Does not apply
1 Less than 12 m^2
2 12 − 30 m^2
3 31 − 90 m^2
4 91 − 450 m^2
5 More than 450 m^2

69 E *Other workplaces*

0 Does not apply
1 *Defined, mobile workplace inside a plant or workshop* (e. g. crane-driver's cab, cab of a fork-lift)
2 *Defined, mobile workplace outside a plant* (e. g. lorry cab)
3 *Non-defined,* mobile workplace *inside* a plant (e. g. quality-checking, work of an office commissionaire)
4 *Non-defined,* mobile workplace *outside* a plant (e. g. activity of a sales representative, work of a mail carrier)
5 *Multiple defined* workplaces

3 Working environment

> The term "working environment" comprises all physical, organizational, social and economic factors that exercise an influence on the performance of the working person.

3.1 Physical environment

CNO   CC   Description of characteristics

3.1.1 Environmental influences

70   E   *Unfavourable illumination effects*

The following combined codes for stress extent/duration should be used to classify the negative effects caused by inappropriate illumination.

Extent of stress	Duration of stress — Proportion of 8-hour shift		
	$1/10 < t \leqslant 1/3$	$1/3 < t \leqslant 2/3$	$t > 2/3$
An illumination intensity appropriate to the required visual efficiency is fully obtained; no unfavourable contrast		0	
An appropriate intensity of illumination is almost obtained; possible occurrence of unfavourable directions of light incidence, shadow effects and contrasts	1	2	3
An appropriate intensity of illumination is not obtained; unsatisfactory illumination of zones; medium extent of glare effects	2	3	4
Very unsatisfactory intensity of illumination for the required visual task; strong glare effects, flicker and stroboscopic effects	3	4	5

If there are several combinations of stress extent/duration during a shift, the highest-rated observed occurrence should be coded. (Examples of appropriate intensity of illumination for different visual tasks are contained in DIN 5035.)

CNO	CC	Description of characteristics
71	E	*Unfavourable climatic conditions*

The following combined codes for stress extent/duration should be used to classify negative effects caused by climatic factors.

Extent of stress	Duration of stress Proportion of 8-hour shift		
	$1/10 < t \leqslant 1/3$	$1/3 < t \leqslant 2/3$	$t > 2/3$
The individual climatic factors (air temperature, humidity, air flow, radiant heat) are fully adapted to the heaviness of the work and the necessary clothing and allow work under comfortable conditions	0	0	0
The adaptation of individual climate elements, heaviness of work and necessary clothing varies only slightly from a neutral thermal value (e. g. for the combination: average humidity of air/low movement of air/no radiant heat/street clothes — sitting — 22° C air temperature .../walking — 15° C air temperature .../ascending stairs — 5° C air temperature)	1	2	3
The adaptation of individual climate elements, heaviness of the work and necessary clothing varies significantly from a neutral thermal value	2	3	4

CNO	CC	Description of characteristics
71	E	*Unfavourable climatic conditions* — continued

Extent of stress	Duration of stress Proportion of 8-hour shift		
	$1/10 < t \leq 1/3$	$1/3 < t \leq 2/3$	$t > 2/3$
Considering the heaviness of work to be analysed and the required clothing, the climatic conditions of the room are near the upper tolerance limits (This may apply to work with foundry ladles, blast furnaces, heat-treated workpieces etc.)	3	4	5

If there are several combinations of stress extent/duration during a shift, the highest-rated observed occurrence is to be coded.

72	E	*Mechanical vibrations*

The following combined codes for stress extent/duration are to be used to classify vibrations of the entire body or of the upper extremities.

Extent of stress	Duration of stress Proportion of 8-hour shift		
	$1/10 < t \leq 1/3$	$1/3 < t \leq 2/3$	$t > 2/3$
No influence of vibration	0		
Vibrations with small amplitudes and at low frequencies beyond the natural frequency of the internal organs (ranges 4–8 Hz vertical influence or 1–2 Hz horizontal influence)	1	2	3
Medium vibration acceleration	2	3	4
Extreme vibration acceleration at frequencies near the natural frequencies of the internal organs. In particular, total-body vibrations; also vibrations in sitting position without any absorption by joints	3	4	5

When there are several combinations of stress extent/duration during a shift, the highest-rated observed occurrence is to be coded.

CNO	CC	Description of characteristics
73	E	*Noise*

The following combined code for stress extent/duration is to be used to classify the stress caused by noise.

Extent of stress	Duration of stress Proportion of 8-hour shift		
	$1/10 < t \leqslant 1/3$	$1/3 < t \leqslant 2/3$	$t > 2/3$
Sound intensity, frequency and timing of the noise stress at the workplace are — for all tasks performed by the incumbent — without any significant effect		0	
Sound pressure level significantly below admissible sound level	1	2	3
Sound pressure level reaches the admissible sound level (e. g. 70 dB(A): normal or mainly mechanized office work, 85 dB(A): other activities at recently designed workplaces; cf. also ArbStättV)	2	3	4
Sound level above admissible value	3	4	5

When there are several combinations of stress extent/duration during a shift, the highest-rated observed occurrence is to be coded.

| 74 | D | *Low or high pressure* |

Indicate the proportion of *time* during which the incumbent must work under air conditions with an atmospheric pressure of 0.9 — 1.0 or is exposed to pressure variations (e. g. caisson work, work at a high altitude).

CNO	CC	Description of characteristics

75   D   *Chemicals, dust, gas, smoke*

Indicate the proportion of *time* during which the incumbent's work is affected by chemical substances (particularly by poisonous or caustic substances and/or by substances irritating the mucous membrane), vapours, dust, smoke etc., or is hindered by such substances.

Classify only inconveniences or impediments; dangerous effects are to be coded under characteristics 80 and 81.

76   D   *Inconvenience due to bad odours*

Indicate the proportion of *time* during which the incumbent is exposed to bad odours.

77   D   *Dirty or wet working environment*

Indicate the proportion of *time* during which the incumbent must work in a dirty environment where he risks getting himself or his clothes smeared or covered with dirt, e. g. car repair shops, coal-mines, painting work, work at lathes or milling machines.

78   D   *Atmospheric influences*

Indicate the proportion of *time* during which the incumbent works out-of-doors and is exposed to atmospheric influences.

79   D   *Impairment by protective clothes, protective devices and safety guards, protective agents*

Indicate the proportion of *time* during which the incumbent is hindered in his freedom of motion and his possibilities of development by protective clothes, protective devices (e. g. safety glasses, insulating gloves, protective mask, ear flaps, safety shoes), safety guards (e. g. covering of moving machine parts, protective screen) and other protective agents (e. g. chemical skin-protectives).

3.1.2 Work hazards or risk of occupational disease

80   E   *Work hazards*

       0 Does not apply
       1 Very minor
       2 Minor
       3 Average
       4 High
       5 Extreme

Answer the following questions considering the extent to which danger is involved in the incumbent's work. For this purpose consider possible dangers due to:

CNO	CC	Description of characteristics

- released energies
- moving work instruments and materials
- exposed workplaces

Examples:
>Danger of stab wounds, bruises, cuts, burns; danger caused by electricity, intoxication, radioactivity, explosion

Ask yourself the following basic questions:
- Are operating and protective devices functionally coupled to each other?
- Are outsiders also protected?
- Are there any new hazards caused by protective measures?
- Is protection ensured even during adjusting and servicing activities?

0
| Assembling
| Operating machine tools
| Mining, stone-breaking
| Preparing blasts
| Decontaminating, extinguishing

81	E	*Occurrence of occupational diseases*

0 Does not apply
1 Very minor
2 Minor
3 Average
4 High
5 Extremely high

Indicate the *probability* of the occurrence of occupational diseases.

This item includes:
- Diseases caused by *chemical influence* (by metals or metalloids, asphyxiant gases, solvents, pesticides etc.)
- Diseases caused by *physical influence* (by chronic mechanical influences, compressed air, noise, heat radiation, ionizing radiation)
- Diseases caused by *infective organisms* or parasites
- Diseases of the lungs or bronchia
- Skin diseases

An *occupational disease* is a long-lasting injury to health that, according to medical findings, is caused by extraordinary conditions to which some people are exposed in a higher degree than others due to their occupation.

CNO	CC		Description of characteristics

3.2 Organizational and social environment

3.2.1 Organization of working time

82	A	*Regularity of working hours*

For the workers there is a timetable that determines the regular daily working time or a regular rotation of shifts.

83	E	*Shift system*

1 Single-shift work in the daytime
2 Work in two shifts
3 Work in three shifts
4 Other shift system

Use the above code to characterize the shift system that normally applies to the incumbent's work.

84	A	*Night-work*

This term applies to all activities that must be — permanently or occasionally — executed during the night, i. e. at least half of the working time is between 22.00 hrs. and 06.00 hrs.

85	E	*Alternating shift system*

0 Does not apply
1 Monthly or more infrequently
2 More than once per month
3 Weekly
4 More than once per week
5 Daily

Use the code above to classify the frequency of change between day and night for the respective job; e. g. for alternating day/night shift the frequency is "daily", for one week day shifts and one week night shifts "weekly".

86	A	*Continuous shift system including weekends*

The shift scheme is *continued over weekends and on holidays*, so that weekend work — i. e. from Friday 24.00 hrs. to Sunday 24.00 hrs. — will be required; weekend work in the form of extra overtime should not be considered.

87	A	*Continuity of work*

Activity is not dependent on season or changes of production, and can be performed over the whole year without change.

88	E	*Starting time in the morning*

0 Does not apply
1 After 08.00 hrs.

CNO	CC	Description of characteristics

        2 After 07.00 hrs. and up to 08.00 hrs.
        3 After 06.00 hrs. and up to 07.00 hrs.
        4. After 05.00 hrs. and up to 06.00 hrs.
        5 05.00 hrs. and earlier

For multi-shift work: Beginning of earliest shift
For flexible working hours: Usual starting time

89   E   *Finishing time in the afternoon*

        0 Does not apply
        1 Up to and including 16.00 hrs.
        2 After 16.00 hrs. and up to 18.00 hrs.
        3 After 18.00 hrs. and up to 20.00 hrs.
        4 After 20.00 hrs. and up to 22.00 hrs.
        5 After 22.00 hrs.

For multi-shift work: End of latest shift
For flexible working hours: Usual finishing time

If the incumbent works almost regular overtime, this fact must be taken into consideration.

90   A   *Flexible working hours*

91   E   *Time additional to regular working hours*
             (overtime hours per week)

        0 Does not apply
        1 Up to and including 2 hours
        2 More than 2 and up to 4 hours
        3 More than 4 and up to 6 hours
        4 More than 6 and up to 9 hours
        5 More than 9 hours

How many overtime hours or other activities outside regular working hours are necessary on average? It is advisable to base calculations of the average number of overtime hours per week on the total working hours per year (if possible) to equalize seasonal variations.

92   E   *Total number of working hours per week*

        1 Up to and including 21
        2 More than 21 and up to 25
        3 More than 25 and up to 30
        4 More than 30 and up to 40
        5 More than 40

How many hours does the incumbent usually work per week *(excluding overtime hours)*?

CNO	CC	Description of characteristics

93 E *Number of longer breaks*
   0 Does not apply
   1 No break provided (e. g. part-time work)
   2 *One* break provided
   3 *Two* breaks provided
   4 *More than two* breaks provided

   Indicate number and duration of longer breaks (⩾ 5 min).

   Consider besides the contractual recreation times also those breaks that are customarily allowed in the organization.

94 E *Total duration of break time*
   1 More than 90 min
   2 61–90 min
   3 31–60 min
   4 16–30 min
   5 15 min and less

   If the total break time *depends on shift type*, the *shortest total break time* should be classified.

95 E *Compulsory breaks due to interference times*
   0 Does not apply
   1 1 – 2 compulsory breaks/8-hour shift
   2 3 – 8 compulsory breaks/8-hour shift
   3 1 – 6 compulsory breaks/hour
   4 7 – 12 compulsory breaks/hour
   5 More than 12 compulsory breaks/hour

   e. g. waiting time proportions of basic time

96 E *Total duration of compulsory breaks due to machine-down times*
   0 Does not apply
   1 Up to and including 5 min/8-hour shift
   2 6–15 min/8-hour shift
   3 16–30 min/8-hour shift
   4 31–60 min/8-hour shift
   5 More than 60 min/8-hour shift

   Ask the incumbent or his superior for the total duration of compulsory breaks caused by machine-down times (related to an 8-hour shift).

CNO	CC	Description of characteristics

3.2.2 Position of the job within the sequence of operations

| 97 | E | *Sequence of operations* |

   0 Does not apply (no visible sequence of operations)
   1 Similar work systems are grouped together *in the same locality*, e. g. typists' room, card punch room; in the processing industry, so-called workshop structure, e. g. lathe shop, milling shop
   2 Work following the *flow principle* (the individual work systems are arranged according to the work process, e. g. a file that is processed progressively from workplace to workplace; in industrial production, both series and flow production)
   3 Work with *automatically controlled sequence of operations*
       (Human influence is restricted to adjusting, charging or discharging activities; e. g. work at filling machines, press lines, fully automatic transfer lines)
   4 Work following *other principles of production* (applies — contrary to the case with stationary work systems of levels 1–3 — to all *mobile* work systems), e. g. site assembly in building construction, plant construction, maintenance activities on stationary work instruments, movable production (road or railway construction) for agricultural and mining work

   Define the man-at-work system according to its position within the work process; activities that are not part of a given work process should not be considered.

| 98 | A | *Individual work* |

   (the work is performed by *one* working person)
   Example of individual work: Working at a lathe, operating a travelling crane, driving a lorry
   Contrary: Group work (several persons — two or more — work simultaneously on one work object)

| 99 | A | *Single-position work* |

   (the work tasks are related to one workplace)
   Contrary: Multi-position work (the job incumbent controls or operates several machines)

| 100 | E | *Work structure* |

   1 Very minor structural restraints
       (the incumbent can determine the organization of his work almost freely, e. g. activities of free-lance artists)
   2 Very minor structural restraints
       (the incumbent has a contractual personal freedom of action to organize his work within the general requirement of the job; e. g. work of a sales representative or a scientist)

53

CNO	CC	Description of characteristics

    3 Average structural restraints
　　　(e. g. activities of teachers, employees in administrative work)
    4 Strong structural restraints
　　　(the incumbent must perform a largely predetermined sequence of tasks)
    5 Very strong structural restraints
　　　(the sequence of tasks is precisely predetermined, e. g. activities following a checklist for aeroplane maintenance, assembling component parts)

Use this characteristic to classify the *structural restraints* of the incumbent's work.

3.2.3 Position of the job within the organizational hierarchy

101   E   *Total number of employees for whom the incumbent is responsible*

    0 Does not apply
    1 10 or fewer employees
    2 11−50 employees
    3 51−250 employees
    4 251−750 employees
    5 751 and more employees

Indicate the number of persons for whom the incumbent is *directly* and *indirectly* responsible, using the above code; e. g. the director of a limited company is *indirectly* responsible for all employees of the company, the director of a branch office is responsible for the employees of the branch, a foreman bears direct responsibility for the employees he supervises.

102   E   *Number of supervisors subordinate to the incumbent for whom he is responsible and to whom he has authority to give instructions*

    0 Does not apply
    1 1 or 2 supervisors
    2 3 or 4 supervisors
    3 5 or 6 supervisors
    4 7 or 8 supervisors
    5 9 and more supervisors

Indicate the number of *supervisors* who are directly *subordinate* to the incumbent of the position to be analysed, using the corresponding code. Supervisors are persons who have responsibility and authority to give instructions to a group of employees.

This characteristic applies to many persons of the middle or high managerial levels, but also to owners of small enterprises, or to persons who delegate their authority to give instructions to others.

CNO	CC	Description of characteristics

103   E   *Exercising authority over persons who have no managerial functions*
- 0 Does not apply
- 1 1–4 employees
- 2 5–8 employees
- 3 9–12 employees
- 4 13–16 employees
- 5 17 and more employees

Using the above code, indicate the number of people who are *directly* instructed by the incumbent. Only those persons should be considered who do production work or perform services and who do not supervise or instruct other persons. This category includes most supervisors on the lower managerial level, e. g. foremen, group leaders, masters in car repair shops, crew leaders, chief pharmacists in larger pharmacies.

104   A   *Exercising staff functions*

The incumbent is a technical advisor; on the basis of certain professional qualifications, he or she provides technical assistance to managers in order to aid them in the decision-making process. The incumbent has no proper instructing authority but bears responsibility. Examples of technical staffs are those of law departments, marketing research units, statistics departments.

105   S   *Responsibility and instructional authority towards persons not employed by the incumbent's organization*

Consider the extent of responsibility and authority, e. g. of teacher towards his pupils, of a physician towards his patients.

Do not estimate the responsibility the incumbent has in relation to colleagues and co-workers.

106   E   *Type of instruction given to the incumbent*
- 0 No instruction concerning type and execution of tasks is required because of the very low level of work complexity and because of its highly repetitive nature
- 1 Precise instruction is given prior to execution of a task
  (e. g. work of an unskilled construction worker)
- 2 Precise instructions concerning the task and general instructions concerning the execution are given
  (the incumbent is precisely instructed about the type of the work to be performed; execution requires only general tuition, e. g. work of an assembly worker, draughtsman, programmer)
- 3 Instructions *exclusively* concerning the task are given
  (the incumbent is precisely instructed about the type of work to be done; the way the work is executed is left to the incumbent; e. g. work of a departmental head or works manager)

CNO	CC	Description of characteristics

4 Very few task instructions are given
(the incumbent is generally instructed concerning his task and the limits within which very complex or extensive tasks are to be performed; e. g. activity of a scientist in research or a manager of a subsidiary company)

5 No task instructions are given
(the incumbent does free-lance work or is self-employed; e. g. work of a practitioner, owner of a shop or enterprise)

Using the above code, indicate the type of work or task instruction usually given to the incumbent.

107   E   *Check of the quality of the work output or product*

Classify the type of check performed on the quality of the work output, using the corresponding code.

0 Does not apply
The incumbent cannot exert influence on the quality of the work output (e. g. machine feeding)

1 The incumbent can influence the quality of the work output but there is *no check* performed; the acceptability of the work product is considered later in the work process

2 *The incumbent checks the work output himself*

3 The work output is checked *within a given group or department* by the *incumbent's colleagues*

4 Execution of the task and the work output *are completely controlled or checked by a supervising person in the given department*

5 The work output is checked by someone *outside the department* in which the incumbent works

If — in the opinion of the analyst — several of these code levels are applicable, the highest-rated observed step is to be coded.

108   E   *Type of managerial functions of the incumbent*

0 Does not apply

1 Direct management
(the incumbent instructs the co-workers for whom he is directly responsible, explains methods of work and maintains frequent contact with his co-workers for the purposes of instruction. That means that he exercises direct influence on the working behaviour of his co-workers; e. g. work of a foreman)

2 Supervisory management
(the incumbent directs and manages his co-workers in a general manner; he gives not detailed but general instructions and allows a considerable freedom for choosing working methods and for work planning; e. g. work of a master mechanic)

CNO	CC	Description of characteristics

    3 Senior management
(the incumbent directs and co-ordinates the activities of *several* working groups. Each working group has its own group leader. The directing or managing functions of the incumbent therefore consist mainly in co-ordinating the work of these leaders; e. g. work of a departmental head)

    4 Management of a complete substantial part of an organization
(the incumbent is fully responsible for a large part of an organization, e. g. for a branch establishment)

    5 Management of an entire organization
(the incumbent is responsible for the whole organization; he directs and manages it independently)

Using the corresponding code, indicate the type and form of the managerial function of the *incumbent towards the employees for whom he is responsible*.

Note that this characteristic applies only when the directing or managing tasks make up an important part of the incumbent's activities.

109	E	*Extent of responsibility for casualties*

    0 Does not apply

    1 Very restricted
(the incumbent bears very little responsibility for the safety of other persons, e. g. when he operates only small hand tools or machines that are not inherently dangerous)

    2 Restricted
(the incumbent is responsible for the security of others only within very narrow limits; e. g. work on punching presses, lathes or other industrial machines)

    3 Average
(the incumbent must make sure that other persons are not injured, e. g. when operating cranes or driving motor cars)

    4 Significant
(the incumbent must be thorough and make sure, on a continuous basis, that serious injuries are avoided, e. g. when handling dangerous chemicals or explosives)

    5 Very significant
(the security of other persons depends mainly on the correct behaviour of the incumbent; e. g. flying a passenger plane, performing surgical operations)

Using the corresponding code, indicate to which extent the incumbent must be careful and make efforts *to prevent other persons from being injured or damaged*. Only immediate damage or injuries that could be caused directly by incorrect behaviour of the incumbent should be taken into consideration.

CNO	CC	Description of characteristics

110   E   *Responsibility for preventing damage to materials and deterioration in quality*
- 0 Does not apply
- 1 Very restricted
- 2 Restricted
- 3 Average
- 4 High
- 5 Very high

Indicate the extent to which incorrect behaviour by the incumbent may lead to *material damage*, e. g. when handling high-quality instruments or machines, precious objects.

Grading should be carried out with regard to

    Possibility of damage
    Probability of damage
    Extent of damage

111   E   *Responsibility for time losses*
- 0 Does not apply
- 1 Very restricted
- 2 Restricted
- 3 Average
- 4 High
- 5 Extreme

Incorrect behaviour of the incumbent may cause *time losses in the working process*, e. g. delays due to inappropriate handling of devices or machines, delays due to inappropriate processing of files, which may lead to time losses for persons or in processes.

112   E   *Responsibility for abstract values*
- 0 Does not apply
- 1 Very restricted
- 2 Restricted
- 3 Average
- 4 High
- 5 Extreme

Indicate the extent of the incumbent's responsibility for abstract values e. g. the reputation of the organization in which the incumbent is employed

3.2.4 Position of the activities within the communication system

This section sets out the miscellaneous persons or groups of persons with whom the incumbent must maintain contact *to be able to fulfil his work assignments*.

CNO	CC	Description of characteristics
		Indicate not only contacts with members of the same organization or enterprise but also with representatives of other organizations if these contacts are *indispensable for the work.*
113	F	*Contacts with persons on higher or highest managerial levels* Directors general of enterprises, members of the government, presidents of trade unions or other big corporations
114	F	*Contacts with persons on subordinate managerial levels* This question is concerned with all necessary contacts with the "subordinate (middle) management", even if slight differences in hierarchy may be neglected in this case; departmental heads and principal departmental heads are both classified under this item, as are company managers, branch directors of banks etc.
115	F	*Contacts with lower supervisory personnel* Contacts with persons who are directly responsible for a working group, e. g. foremen, crew leaders, senior clerks
116	F	*Contacts with persons who have no managerial functions* Industrial workers, craftsmen, financial auditors without managerial functions etc.
117	F	*Conflicts that may originate from the incumbent's position in the organization* Indicate the *frequency of conflicts* that may result from activities performed by the incumbent on account of his functions, e. g. work-related conflicts with colleagues, transfer of co-workers to other positions, refusal of applications for leave, dismissals, check of work output or determination of rest allowances.
118	F	*Conflicts that may originate from contacts with persons not employed by the incumbent's organization* Indicate the *frequency of conflicts* in which the incumbent may be involved and which may result from incompatibility of interests, e. g. committing somebody to a mental institution against the will of that person, taking care of ex-convicts, conflicts arising in customer advisory service work or in the work of police officers or park police.
119	F	*Stress that may originate from the incumbent's conflicts with social standards* Indicate the frequency of possible conflicts with social standards; e. g. a judge whose decisions are bound to the law, a business man negotiating price agreements within a contract.

| CNO | CC | | Description of characteristics |

3.3 Principles and methods of remuneration

3.3.1 Determination of the basis of remuneration

3.3.1.1 Legal basis of remuneration

Indicate any regulations or contractual provisions that form the legal basis of the remuneration for the position to be analysed. Note that possibly none of the following items or *several of them may apply*, e. g. both industrial agreement and employment agreement.

120	A	*By statute*

e. g. civil service salary; fee schedules for lawyers, physicians etc. are also included. Laws that influence the remuneration indirectly (e. g. laws providing for continued wage payment in the event of illness) should not be classified.

121	A	*By industrial agreement*

Wages are fixed by a contractual regulation between the parties to an industrial agreement, i. e. the employers' association or the individual employers on the one hand and the employees' association (trade union) on the other hand.

122	A	*By employment agreement*

Wages are fixed by contract between the worker representatives and the employer.

123	A	*By individual labour contract*

Wages are fixed by contract directly between employer and individual employee.

3.3.1.2 Grading principles

124	A	*Grading by the educational level of the incumbent*

e. g. public employment, craft

125	A	*Grading by the work demands*

This grading is based on an analysis of the demands existing in the incumbent's man-at-work system.

126	A	*Grading by merit*

This grading depends on the incumbent's merit and is based on measurements or on the results of merit ratings.

127	A	*Other grading methods*

Grading by marital status, but not by age or seniority (= years of service in the firm); cf. questions 132 and 133.

CNO	CC	Description of characteristics

3.3.1.3 Mode of payment

| 128 | E | *Payment of wages and salaries* |

        0 Does not apply
        1 Daily
        2 Weekly
        3 Monthly
        4 Annually
        5 Other method of payment (e. g. irregularly paid commission)

Using the above code, indicate the mode of wage or salary payment.

Indicate in the following which method is used to determine the remuneration for the analysed position.

3.3.2 Determination of the remuneration method

| 129 | A | *Summary job evaluation* |

The general difficulty of the work task is evaluated: the task to be evaluated is, e. g., compared to definitions of wage groups, to short definitions of standardized activities and their respective gradings (index of activities) or to already defined reference examples and their gradings.

| 130 | A | *Analytical job evaluation* |

The evaluation of the difficulty of a task is performed according to the individual types of demand: skill, responsibility, stress, environmental influences, which may be described even more precisely; the extent of each type of job demand may be classified in terms like low, average, high; or the extent of job demands may be assigned by comparison with already described reference examples of jobs and their associated gradings.

| 131 | E | *Number of wage or salary groups* |

        0 No groups
        1 Up to 3 groups
        2 4–7 groups
        3 8–11 groups
        4 12–15 groups
        5 More than 15 groups

Using the corresponding code, indicate *whether* and *how many* wage and salary groups are fixed by law (e. g. in salary or fee schedules) or by contract (e. g. in industrial agreements and/or employment agreements).

| 132 | E | *Remuneration by years of age* |

        0 No gradation by age
        1 Maximum reached by the age of 20

CNO	CC	Description of characteristics
		2 Maximum reached between 21 and 30 3 Maximum reached between 31 and 40 4 Maximum reached between 41 and 50 5 Maximum reached after the age of 50 If *remuneration is exclusively by years of age*, use the above code to classify the age group in which the *maximum wage* is earned.
133	E	*Remuneration by seniority (years of service in the firm)* 0 No gradation by seniority 1 Maximum reached after 1 − 5 years of service 2 Maximum reached after 6 − 10 years of service 3 Maximum reached after 11 − 15 years of service 4 Maximum reached after 16 − 20 years of service 5 Maximum reached after more than 20 years of service If *remuneration is exclusively by seniority* (years of service), use the above code to classify the seniority group (years) in which the *maximum wage* is earned.
134	A	*Remuneration by other methods of gradation* e. g. mixed forms of gradation, by seniority and by years of age
135	A	*Piecework* The work output is defined in terms of money.
136	A	*Premium bonus* The performance of the incumbent is determined in terms of quantity per unit of time or by the ratio "allowed time: quantity".
137	A	*Single-factor scheme* The bonus is based on a single factor, e. g. entirely on the basis of quantity.
138	A	*Multi-factor scheme* The bonus is based on several factors, e. g. combined quantity and quality of material yield.
139	A	*Other remuneration methods based on merit* e. g. standardized wages, controlled time wages, task wages, programme wages, contractual wages
140	A	*Time wage with systematic merit evaluation* The total wage is determined by merit *evaluation* or *comparison* but not by measurements. The evaluation may be based on miscellaneous factors such as quality of work or willingness to work
141	A	*Financial rewards given for shift work, work on Saturdays, Sundays and holidays and for excess work*

CNO	CC	Description of characteristics
142	A	*Benefits given for poor working conditions*
		e. g. extra payment for noise or dirty work
143	A	*Voluntary extras*
		e. g. subsidies for travel expenses, *additional* holiday pay, allowances for foremen

Part B — Task Analysis

CNO	CC	Description of characteristics

1 Tasks mainly related to material work objects

144	S	*Adjusting/preparing the work object*

e. g. adjusting an automatic screw machine, preparing assembly parts

145	S	*Transporting work objects or work instruments*

e. g. transporting tools from tool crib to workplace, moving work objects at the workplace

146	S	*Equipping/inserting*

Inserting objects into prepared bores, guides etc., e. g. for equipping printed circuit boards, performing very simple assembly operations at a conveyor belt etc.

Pay attention to the differences in the work elements involved in "feeding" (which *does not* imply a systematic insertion of a given part) and "assembling" (which applies only to complex assembly operations).

147	S	*Processing*

The work object is processed either manually or by means of tools.

148	S	*Adjusting/setting-up/measuring*

Setting up work instruments or work objects to predetermined values, e. g. adjusting a carburettor, balancing an electric circuit

149	S	*Operating/controlling*

Operating machines and technical equipment, controlling production processes, actuating railway switches, operating a crane, driving a car etc.

150	S	*Assembling/disassembling*

Constructional parts are assembled to form a unit either by hand or by means of tools or they are disassembled.

Do not consider work elements that are already classified under characteristic 146.

151	S	*Installing/arranging*

Objects, materials etc. are manually placed in a defined position or arrangement, e. g. arranging books in a library, designing a shop-window, filling up shelves.

Pay attention to the fact that this characteristic should not be used to classify aspects that have already been considered.

152	S	*Feeding/taking out*

Stock is fed into machines or production plants (throwing in, pouring in, inserting) or taken out.

CNO	CC	Description of characteristics
		When classifying work under this characteristic, pay attention to the fact that the material is neither guided nor directed.
153	S	*Performing simple manual operations*

Objects, materials, animals or human beings are carried or moved by means of simple manipulations with possible use of simple auxiliaries, e. g. cleaning work, storage work in warehouses, loading and unloading of conveyor belts, packing, agricultural activities. Typically this kind of work places relatively low demands on careful work execution.

Use this characteristic for the classification of simple manipulations that have not already been considered and that do not have specific professional requirements.

| 154 | S | *Estimating quality, value or condition* |

Quality, condition and/or value of objects are considered as work information and must be estimated, e. g. in the case of machines, motor cars, buildings, antiques.

The incumbent has a considerable freedom of interpretation and judgment for estimation of facts, circumstances or work results.

| 155 | S | *Checking* |

Check of the quality or quantity according to defined standards.

Pay attention to the fact that this characteristic does *not* apply to judgment performances that are classified under the preceding characteristic!

When making this classification, avoid duplicating judgments that entered into previous classifications.

| 156 | S | *Supervising* |

Continuous observation of machines, plant, traffic situations etc. and possible intervention, e. g. in the control stations of power plants or refineries.

CNO   CC   Description of characteristics

2 Tasks mainly related to information-processing and planning

157   S   *Planning/organizing*
           How important is the planning and organization of the incumbent's work and of the work of other persons? The significance of planning and organization is very low if the work tasks are clearly defined for the incumbent and allow no variations, e. g. in flow-line assembly work. However, planning and organization are of high significance for those responsible for supervision of works or personnel. This implies also assessment of timing; how long will it take until a product is delivered, until a defective machine is repaired, or until a new method of production can be used?

158   S   *Coding/decoding information*
           Coding information or translating coded information back to its original form, e. g. reading Morse code, translating foreign languages, using other coding systems such as shorthand, mathematical symbols, computer languages, replacement part numbers.

159   S   *Transcribing information*
           Registering data for further use, e. g. entering meter readings in a record book, entering stock market transactions in a ledger.

160   S   *Arranging/classifying*
           Arranging objects or facts in thematical groups, e. g. compiling data on a given subject, filing correspondence following a certain schedule, grouping data on the basis of subjects or descriptors.
           Do not consider classifications based on quality factors, which were taken into consideration under characteristic 155.

161   S   *Combining information*
           Combining, synthesizing or integrating data from different sources to establish hypotheses and theories or to combine related information to form a clear and comprehensible concept.
           This characteristic is particularly applicable in the following cases: an economist uses information from various sources to predict future economic conditions; a pilot processes various information during the flight; a judge has to consider miscellaneous data when trying a case.

162   S   *Analysing information or data*
           Information must be broken down into *component parts* for identification of *underlying* principles or facts; e. g. the data contained in a financial report must be analysed; critical data must be analysed when diagnosing mechanical disorders or medical symptoms.

CNO	CC	Description of characteristics
163	S	*Counting*
		How important is the simple counting of processes and events for the work of the incumbent?
164	S	*Using mathematics*
		How significant is the mathematical treatment of work-related facts for the incumbent, e. g. algebraic, geometric or statistical methods or probability theory?

CNO	CC	Description of characteristics

3 Tasks mainly related to other persons

165   S   *Public speaking/representing*

Making speeches or formal presentations before a large audience, e. g. speaking at political meetings, giving sermons in church, presenting new products at fairs.

Pay attention to the fact that instructing/training is classified under characteristic 171.

166   S   *Judging human behaviour*

Estimating movements and work processes, e. g. judgment of a sequence of motions during sport competitions, evaluation of the physical condition of a patient by the physician.

167   S   *Serving/supplying/attending*

Performing services, e.g. giving medical care, waiting on tables in a restaurant.

168   S   *Entertaining*

Entertaining or informing other people in the form of an artistic presentation, e. g. work of a radio or TV presenter or interviewer.

169   S   *Advising*

Classify the significance of advisory activities or recommendations for the execution of work.

The incumbent advises other people in legal, financial, clinical, psychological or other aspects in order to contribute to the solution of problems.

170   S   *Negotiating*

Within the framework of his task the incumbent deals with other persons in order to reach an agreement or solution; e. g. engaging in collective bargaining, performing diplomatic missions.

171   S   *Instructing/training*

Teaching of skills or knowledge in a formal or informal manner, e. g teaching in school, managing, teaching an apprentice a craft.

172   S   *Interviewing*

Conducting interviews with special objectives, e. g. interviewing job applicants.

CNO   CC   Description of characteristics

4 Number and frequency of task repetitions

173   E   *Number of operations within the general work task*
            1 1 – 3 tasks
            2 4 – 6 tasks
            3 7 – 9 tasks
            4 10 – 12 tasks
            5 More than 12 tasks

            The characteristics 144 – 172 covered descriptions of tasks. The relationships between tasks and work activities are not always clear cut.

            For this reason, classify the number of operations by using the above code. The work to be analysed has to be dissected into its operation-related primary and secondary tasks according to the following example:

            Work: Varnishing
            Primary task: Varnishing or priming
            Secondary tasks: Preparing parts
                              Preparing paint
                              Suspending parts for drying
                              Cleaning the instruments used
                              Cleaning the cabin

            This example is thus composed of a total of six operation-related tasks.

174   E   *Cycle time of the operation-related primary task*
            0 Does not apply (no repetition existing or visible)
            1 Cycle time more than 45 min
            2 Cycle time 45 – 5 min
            3 Cycle time 5 – 2 min
            4 Cycle time 2 – 1 min
            5 Cycle time less than 1 min

            Using the above code, classify the duration of work cycles for the primary task (concerning the primary task, see preceding characteristic).

            For activities in the field of primarily non-physical work (particularly supervisory work), it is normally very difficult to determine the cycle time. In such cases the characteristic is to be rated 0.

Part C — Analysis of Work Demands

1 Field of work demands: reception of information

1.1 Dimensions of perception

CNO	CC	Description of characteristics

1.1.1 Visual reception of information

Classify the significance of each following characteristic with regard to the overall extent of information input during execution of the task.

175   S   *Structure and patterns*

Significance of the recognition of *structures* and *patterns* during reception of information (also reading print, particularly manuscripts)

```
    S
0 ┬
  │ Filing
  │ Reading displays
  │ Sewing
  │ Engraving
  │ Carpet-weaving
  │ Surface-checking during production of textiles
  │ Fine drawing
```

176   S   *Colours*

How significant is the perception of *colour differences* during information input?

```
    S
0 ┬
  │ Concreting
  │ Filing
  │ Iron-casting
  │ Inserting printed circuit boards
  │ Car-driving
  │ Colour-checking, e. g. during textile production
```

177   S   *Shape and size*

Significance of the perception of the *shape* (square, round, cylindrical etc.) and of the *size* of work-related objects

```
    S
0 ┬
  │ Filing
  │ Concreting
  │ Work of a goldsmith
  │ Packing
  │ Technical drawing
  │ Sorting to size of agricultural products
  │ Checking coin blanks
```

CNO	CC	Description of characteristics

178   S   *Position*

Significance of recognizing the *position* of work objects and work instruments and estimating the distance between incumbent and object.

$0 \mid^S$

- Car-repairing (position of individual tools in the place where they are stored)
- Inserting printed circuit boards (position of transistors or diodes)
- Crane-driving
- Work of radar controllers

179   S   *Quantity/number*

Estimating the *quantity* of objects including estimation of weight, e. g. estimation of the weight of a beam, of the number of bacteria on a slide, of the quantity of screws in a box

$0 \mid^S$

- Flow-line assembly
- Driving a fork-lift
- Crane-driving (estimation of weight)
- Selling of goods by weight

180   S   *Speed of moving objects*

The *speed* of moving objects (machine elements, vehicles, stock etc.) is considered as work information and must be estimated. Additional estimations, e. g. estimation of process speed, should *not* to be classified under this characteristic.

$0 \mid^S$

- Selling
- Concreting
- Flow-line assembly
- Crane-driving
- Car-driving
- Work of radar controllers

1.1.2 Reception of information via sense of hearing

181   S   *Sound patterns*

Recognition of different sound patterns or sound sequences; identification of Morse code, whistling or horn signals etc. Also recognition of speech, e. g. dictations from tape

CNO	CC	Description of characteristics

S
0
    Filing
    Car-repairing
    Producing lathe work
    Crane-driving
    Ship navigation

182   S   *Sound differences and variations*

Recognition of *different sounds* or *sound variations* in the form of loudness, tone colour, pitch and/or tone quality

S
0
    Machine shop management
    Filing
    Train-driving
    Conducting
    Piano-tuning

183   S   *Directional hearing*

Recognition of the *position* of sounds or tones, e. g. localization of critical sounds in running machines (= recognition of possible defects)

S
0
    Machine shop management
    Ship navigation
    Car-repairing

1.1.3 Reception of information via sense of touch or thermo-sensors of the skin

184   S   *Softness and hardness*

Significance of the recognition of *soft* and *hard* surfaces by sense of touch

S
0
    Machine shop management
    Filing
    Operating dough machines
    Producing pottery-ware

CNO	CC	Description of characteristics

185   S   *Roughness and smoothness*
Significance of the recognition of *rough* or *smooth* surfaces by sense of touch

$0 \overset{C}{\vert}$ Machine shop management
Filing
Operating dough machines
Concreting
Producing lathe work
Surface-checking

Each item should be rated in relation to the overall information input.

186   S   *Climatic stimuli*
Significance of the recognition of *hot*, *cold*, *moist* and *dry* surfaces of work-related objects

$0 \overset{S}{\vert}$ Filing
Supervision of air navigation
Baking
Cooking
Ironing
Forging

1.1.4 Reception of information via sense of smell or sense of taste

187   S   *How significant is the information input via sense of smell or sense of taste?*
(Smell or taste qualities: salty, sweet, sour, perfumed, foul, fruity, spicy, burnt, resinous)

$0 \overset{S}{\vert}$ Machine shop management
Concreting
Assembling
Producing lathe work
Preparing food
Taste- or smell-checking of
coffee, tea, wine, tobacco etc.

CNO	CC	Description of characteristics

1.1.5 Proprioceptive information input

188   S   *How significant is the reception of internal stimuli?*

Internal (proprioceptive) sensations of the body (restoring forces of control elements, length of body limbs, perception of own body motion)

$$0 \begin{array}{l} S \\ \end{array} \left| \begin{array}{l} \text{Machine shop management} \\ \text{Filing} \\ \text{Car-driving (accelerating)} \\ \text{Transporting pieces of furniture} \\ \text{Car-repairing} \\ \text{Precision assembly} \end{array} \right.$$

1.2 Modes of recognition

189   A   *Absolute judgment*
of input information

In the case of *absolute judgments,* classifications are made on the basis of previously learned categories

    e. g. in visual reception of information, only 5 —15 different levels can be differentiated (judgment without use of samples or comparative scales)

190   A   *Relative judgment*
of input information

In the case of *relative judgments,* several signals must be present at the same time to allow *comparison of these signals* by the incumbent (judgment with use of samples and comparative scales).

1.3 Accuracy of the reception of information

191   E   *Required accuracy and precision of information input*
1. Very minor (information input is restricted to very gross details)
2. Minor
3. Average
4. High
5. Extreme

Classify the *accuracy and precision* required during the reception of information to fulfil the task.

CNO	CC	Description of characteristics

Object size, admissible number of errors and tolerance during visual reception of information should be taken into consideration.

```
  E
0 ┬
  │ Cleaning buildings
  │ Concreting
  │ Inserting printed circuit boards
  │ Connecting objects to be measured
  │ Work of a radar controller
  │ Work of a goldsmith
  │ Technical drawing
```

2 Field of work demands: decision

2.1 Complexity of decision

CNO	CC	Description of characteristics

Indicate the *complexity of decisions* typical for the job to be analysed, using the classification code given.

Very minor or minor complexity of decision is usually found in cases where the method of solution or individual work procedures are predetermined. A high level of complexity is present when method of solution and work steps are determined in the course of task execution by the incumbent.

Often the viewpoints of incumbent, colleagues and supervisors are different.

192   E   *Complexity of decision*

1. Very minor
   (an unambiguous relationship exists between reception of information and action. Input information consists of only one signal; e. g. the indicator lamp "processing terminated" lights up ⟶ Switch off machine)

2. Minor
   (an unambiguous relation exists between reception of information and action but there are several signals present as input information; e. g. Typewriter switched on? Paper inserted? Original ready for reading? ⟶ Begin to type)

3. Average
   (several signals must be recognized, one out of a number of alternative actions must be selected; there exist clearly defined relations between reception of information and action)

4. High
   (no precisely defined relation between reception of information and action, e. g. instruction activity)

5. Extreme
   (decision-making requires the formation of strategies, alternative courses of action must be compared)

```
   E
0  ┬
   │ Flow-line assembly
   │ Cleaning work
   │ Concreting
   │ Filing
   │ Secretarial work
   │ Supervision of air navigation
   │ Work management
```

CNO	CC	Description of characteristics

2.2 Urgency of decision

193   E   *Urgency of decision*
- 0 Does not apply
- 1 Very minor
  (work progress depends on decision progress)
- 2 Minor
- 3 Average
- 4 High
- 5 Extreme
  (work progress is independent of the decision progress, e. g. work organized in cycles)

For classification of this item, ask yourself the following basic questions:

How long in advance does the incumbent know about the decisions concerning the primary task of his work?

For how long would it be possible for the incumbent to stop his work without negatively affecting the work process of the entire organization (e. g. group, department or company)?

2.3 Required education

Indicate the type of school education that is *required* for the work.

Pay attention to the fact that the school grade "trade school/commercial school" implies a double qualification. Consider this gradation only under this characteristic, not under the one concerning professional education.

194   E   *General education*
- 0 Does not apply
  (only little or no school education is required, no elementary school grade)
- 1 Elementary school grade
- 2 Leaving certificate
  (General Certificate of Education ordinary level, move to upper form of the secondary school, trade school/commercial school
- 3 Admission to professional school
- 4 Matriculation requirements
  (secondary school diplomas)

CNO	CC	Description of characteristics

$0 \left\{\begin{array}{l} E \\ \\ \end{array}\right.$
Flow-line assembly
Street-cleansing
Locksmith's work·
Filing
Data-gathering
Programming
Work preparation
Machine shop management

Use the code below to classify the education necessary for the incumbent's work.

195 E *Professional education*
  0 Does not apply/only very restricted
    (a training of 14 days is sufficient for the job)
  1 Training
    (up to 1 1/2 years of training)
  2 Completed apprenticeship
    (journeyman, skilled worker, clerk)
  3 Technical school diploma
    (engineer, master craftsman or foreman)
  4 Professional school diploma
  5 University education

$0 \left\{\begin{array}{l} E \\ \\ \end{array}\right.$
Cleaning work
Flow-line assembly
Work of on office commissionaire
Locksmith's work
Work of a dental technician
Machine shop management

196 E *Professional experience required for work execution*
  0 Does not apply
    (no professional experience required)
  1 1 month or less
  2 More than 1 month and up to 6 months
  3 More than 6 months and up to 1 year
  4 More than 1 year and up to 3 years
  5 More than 3 years

CNO	CC	Description of characteristics

```
   E
0 ┬
  │  EDP-console-operating (previously, e. g., periphery operating)
  │  EDP work preparation (previously, e. g., operator)
  │  Navigating a ship (previously sailor)
  │  Works manager (previously, e. g., assistant of works manager)
```

Using the above code, classify the time that the incumbent usually requires to gather experience in previous positions that is necessary for the actual position. This may be the case when an employed pharmacist works 1 or 2 years in a pharmacy to gather the practical knowledge necessary for his future function as independent owner of a pharmacy.

Pay attention to the fact that in this item only the duration of professional experience should be classified, *not* the duration of education parallel to work or that of school or university education: this is accounted for under preceding characteristics.

197	E	*Knowledge of foreign languages*

Using the corresponding code, indicate the extent to which foreign languages are necessary for successful execution of the work task.

0   Does not apply
    (the incumbent needs no knowledge of foreign languages to perform his work)
1   Very minor knowledge
    (understanding of certain technical terms in a foreign language is necessary)
2   Minor knowledge
    (the incumbent must have a basic knowledge of one foreign language; e.g. an information officer who gives simple routine information in English and French)
3   Average knowledge
    (the incumbent must be able to read technical literature)
4   Extensive knowledge
    (conversation and business communications must be carried out in a foreign language)
5   Very extensive knowledge
    (the incumbent must possess a perfect knowledge of foreign languages)

CNO	CC	Description of characteristics

```
     E
0 ─┬─
   │ Concreting
   │ Cleaning work
   │ Filing
   │ Taxi-driving
   │ EDP-operating
   │ Programming
   │ Work of a foreign correspondent
   │ Translating, interpreting
```

198   S   *Required advanced professional training*

The incumbent must enlarge and renew his knowledge in the course of his professional tasks. He must be informed about recent developments related to his job, e. g. the work content is subject to frequent major changes (continuous changes of technology etc.), so that a steady training is not sufficient for the successful fulfilment of tasks that include significant decisions. Do not consider further education that aims at a change of profession or professional promotion.

```
     S
0 ─┬─
   │ Cleaning work
   │ Work of a mechanic
   │ Sales activities
   │ Programming
   │ Work of a physician
```

199   S   *Knowledge of special instructions for action*

Significance of knowledge of, e. g., regulations for the operations of machines or technical plant, instructions for application, instructions for the processing of workpieces

```
     S
0 ─┬─
   │ Cleaning work
   │ Flow-line assembly
   │ Filing
   │ Driving a fork-lift
   │ Operating a boring mill
   │ EDP-console-operating
```

3 Field of job demands: action

3.1 Stress due to postural work (postures/body positions)

> Postural work leads to strain due to the need to maintain a certain body position; there are, however, no forces applied outwardly.

CNO	CC	Description of characteristics
		Using the duration code (D), indicate for how long the incumbent performs his work in one of the described body positions. *Consider also the postures that are adopted during movements of the body (e. g. stooping when pushing a car).* Pay particular attention to postures necessitated by work equipment.
200	D	*Sitting, normal* Typing, assembly or checking activities performed in sitting position
201	D	*Sitting, bent* Equipping printed circuit boards, writing activities, crane-driving
202	D	*Standing, normal* Selling, waiting during machine-idle time, e. g. at machine tools
203	D	*Kneeling, normal* Assembling, e. g. in ventilation ducts *Crouching* Changing a tyre *Standing/stooping* Setting up machine tools
204	D	*Standing considerably stooped* Assembling cable terminal boxes, forming grey iron castings

> Pay attention to the fact that the additional use of body extremities (e. g. *standing, arms above head)* is classified unter the characteristics concerning static work).

CNO   CC   Description of characteristics

3.2 Stress due to static work

> The term *static work* implies a long-term (> 4 sec) muscular effort that does not result in a movement of the body (contrary to dynamic work). Static work is therefore not measurable in the mechanical sense.
>
> During static work, muscular effort can take place not only as a result of the exertion of an external force but also because of the effort required to bear the weight of the bodily extremities.

205   D   *Finger — hand — forearm*

                  Characteristic: Muscular effort without support of body weight
                  Example: Grasping and retaining objects, operating a keyboard

```
     D
0    ┬
     │ Filing
     │ Management
     │ Assembling
     │ Secretarial work
     │ Drilling
```

206   D   *Arm — shoulder — back*

                  Characteristic: Muscular effort with support of body weight
                  Example: Carrying, holding or lifting objects, operating pneumatic hammer etc.

```
     D
0    ┬
     │ Management
     │ Filing
     │ Car-driving
     │ Driving a fork-lift
     │ Shunting rail carriages
```

207   D   *Leg — foot*

                  Characteristic: Muscular effort with support of body weight
                  Example: Machine-operating by means of pedals

```
     D
0    ┬
     │ Machine shop management
     │ Filing
     │ Car-driving
     │ Shunting stock carriages
```

CNO	CC	Description of characteristics

3.3 Stress due to heavy dynamic work

Using the duration code (D), indicate the proportion of the shift time during which the incumbent performs *heavy dynamic work* with use of the regions of the body indicated.

> The term *heavy dynamic work* denotes the effort of large (strong) muscle groups, which always requires an increased consumption of energy.

The classification of heavy dynamic work should also consider body movements like walking, running, climbing, or crawling.

208   D   *Using both arms with support of the muscles of the upper part of the body*

       D
0
       Secretarial work
       Cleaning work
       Coal-mining
       Shovelling sand
       Tapping slag

209   D   *Using both legs with support of the pelvic muscles*
(including walking, climbing, crawling)

       D
0
       Secretarial work
       Flow-line assembly
       Cleaning work
       Work of a mining deputy
       Coal-mining

3.4 Stress due to active light work

Using the duration code (D), indicate the proportion of the shift time during which the incumbent performs *active light work* with use of the regions of the body indicated.

> The term *active light work* denotes the dynamic work of one or several muscle groups the active mass of which is smaller than 1/7 of the overall muscle mass of the body and whose frequency of actuation is higher than 15 exertions per minute.
>
> *Example:* Typing; note that the actual typing is to be considered as active light work, the keeping of the hands in a largely stretched-out position as static work.

CNO	CC	Description of characteristics

210   D   *Using the finger — hand system*
(including use of both hands)

       D
0 ⊤
       | Management
       | Production of lathe work
       | Operating a keyboard
       | Precision assembly
       | Inserting printed circuit boards

211   D   *Using the hand — arm system*
(including the use of both arms without support of the muscles of the upper part of the body)

       D
0 ⊤
       | Management
       | Supervision of air navigation
       | Cleaning work
       | Assembling
       | Producing bicycle tyres

212   D   *Using the foot — leg system*
(including the use of both feet; the use of both legs with participation of the pelvic muscles should be rated as heavy dynamic work)

       D
0 ⊤
       | Management
       | Filing
       | Producing tyre bodies
       | Work at industrial sewing machines

3.5 Application of forces and frequency of motion

213   E   *Application of forces for static work*
     0   Does not apply
     1   Very minor
     2   Minor
     3   Average
     4   High
     5   Very high

Indicate the *extent of force* (*in relation to the region of the body involved*) that is normally applied for *static work* by the incumbent.

CNO	CC	Description of characteristics

```
    E
0   ┬
    │  Repairing watches/clocks
    │  Secretarial work
    │  Welding
    │  Concreting
    │  Transporting moulds
    │  Transporting pieces of furniture
```

214 E *Application of forces for heavy dynamic work*
   0 Does not apply
   1 Very minor
   2 Minor
   3 Average
   4 High
   5 Very high

Indicate the *extent of force* (*in relation to the region* of the body involved) that is normally applied for *heavy dynamic work* by the incumbent.

```
    E
0   ┬
    │  Walking on the ground
    │  Walking on a stubble-field
    │  Climbing (inclined plane, 10°)
    │  Shovelling (load 8 kg, 15 throws/min, throwing distance 1 m)
    │  Climbing a ladder (70 steps/min)
    │  Hewing coal
    │  Working with an axe
```

215 E *Application of forces for active light work*
   0 Does not apply
   1 Very minor
   2 Minor
   3 Average
   4 High
   5 Very high

Indicate the *extent of force* (*in relation to the region of the body involved*) that is normally applied for *active light work*.

```
    E
0   ┬
    │  Secretarial work
    │  Coding data on punched cards/tapes
    │  Inserting printed circuit boards
    │  Production of lathe work
    │  Assembling bicycle/scooter tyres
```

CNO	CC	Description of characteristics
216	E	*Frequency of motion for active light work*

0 Does not apply
1 Very minor
2 Minor
3 Average
4 High
5 Very high

Indicate the *frequency of motion* that is usually required for the incumbent's work.

Consider the relationship to the region of the body used; e. g. high frequencies of motion of

fingers	$\geqslant$ 120 motions/min
hand	$\geqslant$ 90 motions/min
forearm	$\geqslant$ 90 motions/min
upper arm	$\geqslant$ 60 motions/min
foot	$\geqslant$ 60 motions/min
leg	$\geqslant$ 60 motions/min

E
- Crane-driving
- Inserting printed circuit boards
- Coding (punching)
- Typing

References

ArbStättV: Verordnung über Arbeitsstätten

DIN 5035 Beleuchtung
Teil 1: Begriffe und allgemeine Anforderungen
Teil 2: Richtwerte für Arbeitsstätten

Fleishman, E.A.: Taxonomic problems in human performance research, in: Singleton, W.T.; Spurgeon, P. (Hrsg.): Measurement of human resources, London: Taylor & Francis Ltd., 1975, 49-72

Frieling, E.; Hoyos, C. Graf: Fragebogen zur Arbeitsanalyse — FAA, Bern: Hans Huber Verlag 1978

Hoyos, C. Graf: Arbeitspsychologie. Stuttgart: Verlag W. Kohlhammer GmbH, 1974

Jeanneret, P.R.: A study of the job dimensions of "worker-orientated" job variables and of their attribute profiles. Lafayette: Unpublished Doctoral Dissertation, Purdue University, 1969

Kirchner, J.H.: Analyse von Arbeitssystemen. In: Lehmann, G.: Praktische Arbeitsphysiologie, 3. neubearbeitete Auflage, herausgegeben von W. Rohmert und J. Rutenfranz, Thieme, Stuttgart 1982

Landau, K.: Das Arbeitswissenschaftliche Erhebungsverfahren zur Tätigkeitsanalyse — AET. Dissertation am Fachbereich Maschinenbau der TH Darmstadt, 1978

Landau, K.; Luczak, H.; Rohmert, W.: Arbeitswissenschaftlicher Erhebungsbogen zur Tätigkeitsanalyse. In: Rohmert, W.; Rutenfranz, J.: Arbeitswissenschaftliche Beurteilung der Belastung und Beanspruchung an unterschiedlichen industriellen Arbeitsplätzen. Der Bundesminister für Arbeit und Sozialordnung, Bonn 1975

Landau, K.; Luczak, H.; Rohmert, W.: Clusteranalytische Untersuchungen zum arbeitswissenschaftlichen Erhebungsbogen zur Tätigkeitsanalyse — AET. Zeitschrift für Arbeitswissenschaft, 30, 1976, 1, 31-39

Landau, K.; Rohmert, W. (Hrsg.): Fallbeispiele zur Arbeitsanalyse — Ergebnisse zum AET-Einsatz. Bern: Hans Huber Verlag 1981

Landau, K.; Rohmert, W.: AET — a new job analysis method; in: Proc. 1981 Spring Annual Conference & World Productivity Congress, Detroit, May 17-20, 1981, p. 751-760

Landau, K.; Rohmert, W.; Reus, J.: Das Arbeitswissenschaftliche Erhebungsverfahren zur Tätigkeitsanalyse (AET) — Auswerteservice, Bern: Hans Huber Verlag 1980

Luczak, H.; Landau, K.; Rohmert, W.: Faktorenanalytische Untersuchungen zum Arbeitswissenschaftlichen Erhebungsbogen zur Tätigkeitsanalyse — AET. Zeitschrift für Arbeitswissenschaft, 30, 1976, 1, 22-30

Luczak, H.; Rohmert, W.; Singer, R.; Rutenfranz, J.: Klassifikation der Arbeitszufriedenheit und des subjektiven Belastungs- und Gesundheitsstatus von Fluglotsen. Int. Arch. Occup. Environ. Hlth., 39, 1977, 1-26

McCormick, W.T.; Mecham, R.C.; Jeanneret, P.R.: The development and background of the Position Analysis Questionnaire. Occupational Research Center, Purdue University, Lafayette 1969

McCormick, E.J.; Mecham, R.C.; Jeanneret, P.R.: Technical manual for the position analysis questionnaire (P.A.G.). Lafayette: PAQ Services, Inc. 1972

REFA (Hrsg.): Methodenlehre des Arbeitsstudiums, Teil 1 — Grundlagen. München: Carl Hanser Verlag, 1978

Rohmert, W.: Aufgaben und Inhalt der Arbeitswissenschaft. Die berufsbildende Schule, 24, 1972, 1, 3-14

Rohmert, W.; Rutenfranz, J.: Arbeitswissenschaftliche Beurteilung der Belastung und Beanspruchung an unterschiedlichen industriellen Arbeitsplätzen. Gutachterliche Stellungnahme, Der Bundesminister für Arbeit und Sozialordnung, Bonn 1975

Rohmert, W.; Landau, K.: Das Arbeitswissenschaftliche Erhebungsverfahren zur Tätigkeitsanalyse (AET). Bern-Stuttgart-Wien: Verlag Hans Huber, 1979

INDEX

Accidents 13
AET (ergonomic job analysis procedure) 15, 18
   training of AET analysts 18
   evaluation of AET data 22
   planning an AET job analysis 18, 19
Behaviour requirements 15
Breaks in worktime 52

Casualty responsibility 57
Climatic conditions 45
Cluster analysis 25, 26
Coding of work characteristics 17, 29
Communication within organisation 58-59
Conditions of work 15, 44-63

Dangerous work 36, 48
Data handling 69
Decision 16, 81-82
Demands on worker 15
Duration code 29, 30

Education 13, 14
Educational requirements 82-85
Environment 10, 15, 16, 44-63
     unpleasant conditions 44-49
Equipment 16
   control equipment 40
   for energy transformation 39
   for information processing 39
   for material transformation 38
   for monitoring 41
   for moving materials 39
   for use on living work objects 39
Evaluation of AET data 22
Exclusive code 29, 30
Extreme conditions 22

Factor analysis 22, 23
Flexible work hours 51
Frequency distribution of work characteristics 22, 23
Frequency of motion in active work 91
Frequency of task repetition 72

Grading 60

Handicapped persons 13
Heavy dynamic work 88, 90
Hours of work 50-52
Human relations tasks 71

Illumination 44
Industrial organisation 13
Industrial psychology 9, 13

Information 37
   processing 39, 69
   reception 75-80
Informatory work 9
Instructions for tasks 55
Interviews 15, 18-20

Job demand analysis 15, 16

Light active work 88-91

Man-at-work system, model 10
Managerial functions 56-57
Material work objects 33-36
   dangerous 36
   sensitive 34
   shape 35
   size 35
   special properties 34
   surface conditions 34
   weight 36
Mental attitude of worker 11
Mental work 37

Noise 47

Observation at the workplace 18-20
Organizational hierarchy 54
Overtime 51

Personal characteristics of worker 18
Personnel management 13
Perception 16, 75-80
Profile analysis 22, 24

Quality control 56
Questionnaire, AET 18

Remuneration 13, 16, 60-63
Repetition of tasks 72
Repetitiveness of work 18, 19
Responsibility 54-58

Scales for classifying work characteristics 17
Seating 41
Sequence of operations 53
Shiftwork 50
Significance code 29, 30
Sitting work posture 25, 86
Sound patterns for information reception 76-77
Standing work posture 25, 86
Static work 87, 89
Stress/strain 10, 11, 12, 17
   physical 86-91
Supervision 54-56

Task analysis 15, 65-72
Training requirements 83-85

Vibration, mechanical 46
Visual reception of information 75-76
Vocational counselling 14

Work design 13

Work object 10, 15, 16, 33
  energy 37
  information 37
  living 37
  manipulations 67-68
  material 33-36
Working posture 25, 86
Workplace characteristics 41-43